VLSI DESIGN FOR MANUFACTURING: YIELD ENHANCEMENT

THE KLUWER INTERNATIONAL SERIES IN ENGINEERING AND COMPUTER SCIENCE

VLSI, COMPUTER ARCHITECTURE AND DIGITAL SIGNAL PROCESSING

Consulting Editor
Jonathan Allen

Other books in the series:

Logic Minimization Algorithms for VLSI Synthesis. R.K. Brayton, G.D. Hachtel, C.T. McMullen, and Alberto Sangiovanni-Vincentelli. ISBN 0-89838-164-9.
Adaptive Filters: Structures, Algorithms, and Applications. M.L. Honig and D.G. Messerschmitt. ISBN 0-89838-163-0.
Introduction to VLSI Silicon Devices: Physics, Technology and Characterization. B. El-Kareh and R.J. Bombard. ISBN 0-89838-210-6.
Latchup in CMOS Technology: The Problem and Its Cure. R.R. Troutman. ISBN 0-89838-215-7.
Digital CMOS Circuit Design. M. Annaratone. ISBN 0-89838-224-6.
The Bounding Approach to VLSI Circuit Simulation. C.A. Zukowski. ISBN 0-89838-176-2.
Multi-Level Simulation for VLSI Design. D.D. Hill and D.R. Coelho. ISBN 0-89838-184-3.
Relaxation Techniques for the Simulation of VLSI Circuits. J. White and A. Sangiovanni-Vincentelli. ISBN 0-89838-186-X.
VLSI CAD Tools and Applications. W. Fichtner and M. Morf, Editors. ISBN 0-89838-193-2.
A VLSI Architecture for Concurrent Data Structures. W.J. Dally. ISBN 0-89838-235-1.
Yield Simulation for Integrated Circuits. D.M.H. Walker. ISBN 0-89838-244-0.
VLSI Specification, Verification and Synthesis. G. Birtwistle and P.A. Subrahmanyam. ISBN 0-89838-246-7.
Fundamentals of Computer-Aided Circuit Simulation. W.J. McCalla. ISBN 0-89838-248-3.
Serial Data Computation. S.G. Smith and P.B. Denyer. ISBN 0-89838-253-X.
Phonologic Parsing in Speech Recognition. K.W. Church. ISBN 0-89838-250-5.
Simulated Annealing for VLSI Design. D.F. Wong, H.W. Leong, and C.L. Liu. ISBN 0-89838-256-4.
Polycrystalline Silicon for Integrated Circuit Applications. T. Kamins. ISBN 0-89838-259-9.
FET Modeling for Circuit Simulation. D. Divekar. ISBN 0-89838-264-5.
VLSI Placement and Global Routing Using Simulated Annealing. C. Sechen. ISBN 0-89838-281-5.
Adaptive Filters and Equalizers. B. Mulgrew, C.F.N. Cowan. ISBN 0-89838-285-8.
Computer-Aided Design and VLSI Device Development, Second Edition. K.M. Cham, S-Y. Oh, J.L. Moll, K. Lee, P. Vande Voorde, D. Chin. ISBN: 0-89838-277-7.
Automatic Speech Recognition. K-F. Lee. ISBN 0-89838-296-3.
Speech Time-Frequency Representations. M.D. Riley. ISBN 0-89838-298-X
A Systolic Array Optimizing Compiler. M.S. Lam. ISBN: 0-89838-300-5.
Algorithms and Techniques for VLSI Layout Synthesis. D. Hill, D. Shugard, J. Fishburn, K. Keutzer. ISBN: 0-89838-301-3.
Switch-Level Timing Simulation of MOS VLSI Circuits. V.B. Rao, D.V. Overhauser, T.N. Trick, I.N. Hajj. ISBN 0-89838-302-1
VLSI for Artificial Intelligence. J.G. Delgado-Frias, W.R. Moore (Editors). ISBN 0-7923-9000-8.
Wafer Level Integrated Systems: Implementation Issues. S.K. Tewksbury. ISBN 0-7923-9006-7
The Annealing Algorithm. R.H.J.M. Otten & L.P.P.P. van Ginneken. ISBN 0-7923-9022-9.
VHDL: Hardware Description and Design. R. Lipsett, C. Schaefer and C. Ussery. ISBN 0-7923-9030-X.
The VHDL Handbook. D. Coelho. ISBN 0-7923-9031-8.
Unified Methods for VLSI Simulation and Test Generation. K.T. Cheng and V.D. Agrawal. ISBN 0-7923-9025-3
ASIC System Design with VHDL: A Paradigm. S.S. Leung and M.A. Shanblatt. ISBN 0-7923-9032-6.
BiCMOS Technology and Applications. A.R. Alvarez (Editor). ISBN 0-7923-9033-4.
Analog VLSI Implementation of Neural Systems. C. Mead and M. Ismail (Editors). ISBN 0-7923-9040-7.
The MIPS-X RISC Microprocessor. P. Chow. ISBN 0-7923-9045-8.
Nonlinear Digital Filters: Principles and Applications. I. Pitas and A.N. Venetsanopoulos. ISBN 0-7923-9049-0.
Algorithmic and Register-Transfer Level Synthesis: The System Architect's Workbench. D.E. Thomas, E.D. Lagnese, R.A. Walker, J.A. Nestor, J.V. Rajan, R.L. Blackburn. ISBN 0-7923-9053-9.

VLSI DESIGN FOR MANUFACTURING: YIELD ENHANCEMENT

by

Stephen W. Director
Wojciech Maly
Andrzej J. Strojwas

Carnegie Mellon University

KLUWER ACADEMIC PUBLISHERS
Boston/Dordrecht/London

Distributors for North America:
Kluwer Academic Publishers
101 Philip Drive
Assinippi Park
Norwell, Massachusetts 02061 USA

Distributors for all other countries:
Kluwer Academic Publishers Group
Distribution Centre
Post Office Box 322
3300 AH Dordrecht, THE NETHERLANDS

Library of Congress Cataloging-in-Publication Data

Director, Stephen W.
VLSI design for manufacturing : yield enhancement / by Stephen W. Director, Wojcilech Maly, Andrzej J. Strojwas.
p. cm. — (Kluwer international series in engineering and computer science. VLSI, computer architecture, and digital signal processing)
Includes bibliographical references.
ISBN 0-7923-9054-7
1. Integrated circuits—Very large scale integration—Design and construction—Data processing. 2. Computer-aided design. I. Maly, W. II. Strojwas, Andrzej J. III. Title. IV. Series.
TK7874.D555 1990
621.39′5—dc20 89-37029
CIP

Printed in the United States of America

To our students

Table of Contents

1. Yield Estimation and Prediction 1

1.1. Introduction 1

1.2. The VLSI Fabrication Process 2

1.3. Disturbances in the IC Manufacturing Process 11

1.3.1. Process Disturbances 11

1.3.2. Process Related Deformations of IC Design 13

1.3.2.1. Geometrical Deformations 13

1.3.2.2. Electrical Deformations 18

1.3.3. General Characteristics of Process Disturbances 20

1.3.4. IC Performance Faults 20

1.4. Measures of Process Efficiency 22

1.4.1. Yield Estimation 23

1.4.2. Yield Prediction 26

1.4.3. Decomposition of the Design Yield Equations 29

1.5. Discussion 31

1.5.1. Relationships Between Manufacturing and Design Yields 31

1.5.2. Examples of Yield Analysis 33

1.5.3. Yield and Production Cost 37

1.6. Overview of the Sequel 40

2. Parametric Yield Maximization 43

2.1. Introduction 43

2.1.1. Definitions of Yield and Design Center 44

2.1.2. Design Centering By Simplicial Approximation 46

2.1.3. Design Centering Procedure 51

2.1.4. Scaling: Inscribing a Hyperellipsoid 55

2.1.5. Illustration of the Basic Method 58

2.2. Design Centering and Worst Case Design with Arbitrary Statistical Distributions 62

2.2.1. Norm Bodies and PDF Norms 62

2.2.2. Generalized Simplicial Approximation 67

2.2.3. Mixed Worst-Case Yield Maximization 70

2.2.4. Tolerance Assignment 76

2.3. Example of Worst Case Design 79

2.4. A Dimension Reduction Procedure 81

2.4.1. Design Centering in a Reduced Space 83
2.4.2. Discussion 85
2.5. Fabrication Based Statistical Design of Monolithic IC's 86
2.5.1. Independently Designable Parameters in Yield Maximization of Monolithic IC's 87
2.5.2. Process Simulation and Yield Maximization 89

3. Statistical Process Simulation 93

3.1. Introduction 93
3.2. Statistical Process Simulation 94
3.2.1. Methodology 94
3.2.2. Modeling of Process Disturbances 97
3.2.3. Process and Device Models for Statistical Simulation 105
3.2.4. Models of Ion Implantation 106
3.2.5. MOS Transistor Model 108
3.2.6. Structure of the Simulator 109
3.3. Tuning of Process Simulator with PROMETHEUS 112
3.3.1. Mathematical Formulation 112
3.3.2. Methodology of Solution 115
3.4. The Process Engineer's Workbench 116
3.4.1. Process and Device Simulation 117
3.4.2. User Interaction-Process Synthesis 119
3.4.3. Internal Data Structures 121
3.4.4. User Interaction - Compiled Simulation 124
3.4.5. Extensions 128

4. Statistical Analysis 129

4.1. Statistical Timing Simulation 129
4.1.1. Overview 130
4.1.2. Our Approach 131
4.1.2.1. Timing reevaluation 136
4.1.3. Characterization 137
4.1.4. Delay decomposition 140
4.1.5. Nominal Simulation 143
4.1.6. Statistical Simulation 147
4.2. An Improved Worst-Case Analysis Procedure 152
4.2.1. Worst-Case Analysis Methodology 153
4.2.1.1. Algorithm for Worst-Case Analysis 156
4.2.2. A Software Package for Worst-Case Analysis 156

4.2.3. Examples 158
4.3. Optimal Device and Cell Design Using FABRICS 166
4.3.1. Proposed Methodology 167
4.3.2. Description of Experiment 168
4.3.3. Building the Regression Model 171
4.3.4. Performance Optimization 172

5. Functional Yield **175**

5.1. Introduction 175
5.2. Basic Characteristics of Spot Defects 176
5.2.1. Defect Mechanisms 177
5.2.2. Defect Spatial Distribution 177
5.2.3. Distribution of Defect Radii 184
5.2.4. Distribution of Defect Radii Within Layer 186
5.3. Yield Modeling Using Virtual Layout 189
5.3.1. Critical Area 190
5.3.2. Spot Defect Related Yield Losses 191
5.3.3. Yield Losses Due to Lateral Process Deformations 195
5.3.4. Critical Area Computation Using Virtual Layout 198
5.3.5. Examples of Application of the Virtual Layout Method 200
5.4. Monte Carlo Approach to Functional Yield Prediction 203
5.4.1. The VLASIC Yield Simulator 204
5.4.2. Fault Analysis 205
5.4.3. VLASIC Implementation and Summarizing Discussion 212
5.5. Yield Computations for VLSI Cell 213
5.5.1. Probability of Failure (POF) for Simple Layout Patterns 213
5.5.2. POF for Macrocells 220
5.5.3. Implementation 228

6. Computer-Aided Manufacturing **229**

6.1. Motivation 229
6.2. Overview of the CMU-CAM System 231
6.3. Statistical Process Control: The Unified Framework 233
6.3.1. Profit Function 236
6.4. CMU-CAM Software System 238
6.4.1. Decomposition 238
6.4.2. Modeling for Process Control 242
6.4.3. Statistical Quality Control 245
6.4.4. Acceptance and Rejection Criteria 249

6.4.5. Feed Forward Control 256
6.5. Computational Examples 260
6.5.1. Yield Enhancement 265
6.6. Conclusions 267
References 269
Index **285**

Preface

One of the keys to success in the IC industry is getting a new product to market in a timely fashion and being able to produce that product with sufficient yield to be profitable. There are two ways to increase yield: by improving the control of the manufacturing process and by designing the process and the circuits in such a way as to minimize the effect of the inherent variations of the process on performance. The latter is typically referred to as "design for manufacture" or "statistical design". As device sizes continue to shrink, the effects of the inherent fluctuations in the IC fabrication process will have an even more obvious effect on circuit performance. And design for manufacture will increase in importance. We have been working in the area of statistically based computer aided design for more than 13 years. During the last decade we have been working with each other, and individually with our students, to develop methods and CAD tools that can be used to improve yield during the design and manufacturing phases of IC realization. This effort has resulted in a large number of publications that have appeared in a variety of journals and conference proceedings. Thus our motivation in writing this book is to put, in one place, a description of our approach to IC yield enhancement. While the work that is contained in this book has appeared in the open literature, we have attempted to use a consistent notation throughout this book. (Despite our best efforts thought there are times that inconsistency has crept in and we ask the readers patience.) We have also decided to organize the material by topic, rather than in chronological order. Thus it is not necessarily true that the material presented in earlier chapters was actually developed prior to the material presented in later chapters.

We begin in Chapter 1 by presenting a uniform framework of viewing yield related problems in IC manufacture. The concepts of parametric and functional yields are introduced and a number of important definitions and assumptions are presented. Chapter 2 discusses the parametric yield maximization problem and introduces the simplicial approximation method. This method will also play a role in a computer aided manufacturing system presented in Chapter 6. The statistically based process and device simulator FABRICS is described in Chapter 3. This simulator, which can be used to predict parametric yield, plays a central role in much of our work. Several applications of FABRICS are discussed in Chapter 4. We turn to functional yield in Chapter 5. Here we consider in some detail the causes of functional yield loss and three techniques for predicting functional yield. Finally, in Chapter 6 we describe a comprehensive approach to computer-aided manufacturing that is based upon the concepts introduced in the first five chapters.

The work that is described in this book has been influenced greatly by the efforts of a large number of our graduate students. We are indebted to them and gratefully acknowledge them: Jacques Benkoski, Ihao Chen, Marko Chew, Janusz Deszczka, Demitri Giannopolus, Tomek Gutt, Molly Johnson, Pat Kager, John Kibarian, Michael Lightner, K. K. Low, P. K. Mozumder, Sani Nassif, Peter Odryna, Rahul Razdan, Mary Jo Saccamango, C. R. Shyamsundar, Ryszard Skrocki, Costas Spanos, Xiaowei Tian, Mike Trick, Luis Vidigal, Hank Walker, and Chi Min Yuan. It is to them that we dedicate this book.

We would also like to acknowledge that some of the material in this book is also based upon work we have done with some of our other colleagues: Gary Hachtel, Bob Brayton, Jan Koszur, Wiesiek Kuzmicz, Bozena Lesinska, Zhi-Jie Li, Mike Thomas, Randy Hughes and Alfred Swit. Our thanks also go to Ms. Julie Gawdyda who helped in the preparation of this manuscript.

Finally, we would like to acknowledge the financial support of the Semiconductor Research Corporation, the National Science Foundation, IBM, AT&T Bell Laboratories, Harris Semiconductor, Texas Instruments, General Electric, Digital Equipment Corporation, and National Semiconductor that has allowed us to carry out much of this work. We are especially grateful to our industrial friends for providing us with the opportunity to test out many of our ideas in a real world setting.

Chapter 1
Yield Estimation and Prediction[1]

1.1. Introduction

Due to inherent fluctuations in any integrated circuit manufacturing process, the yield (which is nominally viewed as the ratio of the number of chips that perform correctly, i.e., meet all performance specifications, to the number of chips manufactured) is always less than 100%. As the complexity of VLSI chips increases, and the dimensions of VLSI devices decrease, the sensitivity of performance to process fluctuations increases, thus further reducing the manufacturing yield. Since profitability of a manufacturing process is directly related to yield, the search for computer-aided methods for maximizing yield through improved design methods and control of the manufacturing process has intensified dramatically.

Statistical approaches to yield modeling and optimization have been under development for a number of years. For the most part, these methods can be separated into two categories: parametric yield estimation and optimization techniques and catastrophic yield estimation and optimization techniques. In general, parametric yield optimization has been formulated

---

[1]This chapter is based upon the paper "VLSI Yield Prediction and Estimation: A Unified Framework" by W. Maly, A.J. Strojwas and S.W. Director, IEEE Transactions on Computer Aided Design of Integrated Circuits and Systems, CAD-5(1), pp. 114-130, January 1986.

as a tolerance assignment or design centering problem [5], [23], [90], [24], [7], [120], [91], [13], [55], [3], [116]. However, due to simplified assumptions about circuit element characteristics, these approaches have been proven successful only for discrete circuits that have a relatively small number of designable parameters. To overcome the above drawbacks, techniques for yield optimization that employ statistical process simulation techniques [49], [60], [65], [104], [83] have been developed to handle IC manufacturing process related constraints [63], [64], [115], [94], [35], [4], [20]. Methods for catastrophic yield estimation have been based upon such simplifying assumptions as yield loss being due to point defects only [79], [76], [85], [106], [107], [38]. More elaborate models that handle nonuniform defect distribution [129], [106], defect size distribution [87], [9], [66], [68], [31], [32] and the yield for chips with redundant circuitry [109], [110], [73] have also been reported.

All of the above approaches, while appropriate for the particular applications they were designed for, are not sufficient to handle all of the factors that affect the yield of VLSI circuits. In this chapter we develop a unified framework from which improved methods for yield maximization can proceed. We present a general and realistic approach to predicting and estimating the manufacturing yield of VLSI circuits by taking into account all of the physical phenomena that affect manufacturing yield. We begin in the next section by investigating the structure and organization of a typical VLSI fabrication process. A description of the physical causes of yield loss is then presented along with a classification scheme for possible failure modes in VLSI circuits. We then introduce formulae that can be used to estimate yield during the manufacturing process and to predict yield during the design process. Finally, we discuss the cost and profit aspects of VLSI manufacturing and relate them to yield maximization.

1.2. The VLSI Fabrication Process

As a first step towards developing a manufacturing-based approach to CAD we formalize some terms concerning the structure and organization of a fabrication process. In this section we analyze, in general terms, the VLSI manufacturing process organization and environment, and establish relationships between what is normally called the IC design domain and the IC manufacturing domain.

The IC manufacturing process involves a sequence of basic processing steps that are performed on sets of wafers called *lots*. Each wafer may contain as many as several hundred IC chips. A few of the chips on the wafer are special in that they contain *test structures* and are called *test chips* . Test structures are special purpose devices which are designed so that their performance is sensitive to the quality of specific processing steps. Examples of test structures are long contact chains, large capacitors, and arrays of transistors of various geometries.[2]

To facilitate process observability and process control, one or more processing steps may be separated by an inspection and selection step . An inspection step involves a measurement, or sequence of measurements, called *in-line measurements* , performed on the wafer or, in some cases a special test wafer, that does not contain IC's. If one or more in-line measurements fall outside of some range defined by a set of *selection thresholds* , the wafers are considered defective and are discarded. The conditions under which the in-line measurements are performed, defined in terms of such parameters as voltages or currents, are called *test conditions*. We refer to a sequence of processing steps followed by an inspection and selection step as a *manufacturing operation* as shown in Fig. 1-1.

The outcome of a manufacturing operation depends on three major factors: the process controlling parameters , or *control*; the geometry of the fabricated IC, or *layout*, and some randomly changing environmental factors, called *disturbances*. The *control* of a manufacturing operation is the set of parameters which can be manipulated in order to achieve some desired change in the fabricated IC structure. Examples of control parameters are processing equipment settings which determine temperatures, gas pressures, step duration etc. The *layout* of an IC is described as a set of masks where each mask can be viewed as a collection of transparent regions within an opaque region. The *disturbances* are environmental factors that cause variations in the outcome of a manufacturing operation. Such variations are inherent in any manufacturing process and can be characterized in terms of a set of random variables.

---

[2] In some cases, test structures are also placed in scribe lanes, i.e. in the spaces between the chips on a wafer.

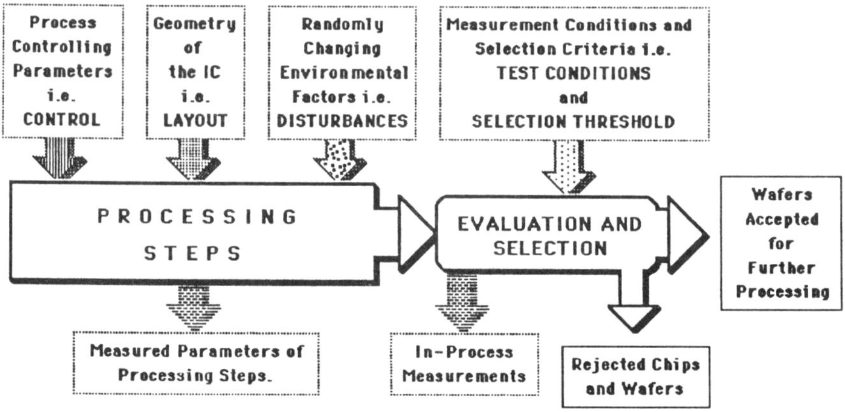

Figure 1-1: Flow of information in a manufacturing operation.

A typical IC fabrication process involves many manufacturing operations, as shown in Fig. 1-2. Observe that the last few manufacturing operations involve probe tests . As shown, probe tests are performed on two different types of chips: regular chips and test chips. Regular chips are classified as passing or failing, and those which pass are forwarded to the assembly and packaging operations. Since test chips are designed to provide information about the process performance for diagnostic purposes, and are not used to select IC chips, test chip measurements are distinguished as a separate branch in the flow diagram. From Fig. 1-2 we see that the IC manufacturing process is composed of a wafer fabrication part, a testing part and an assembly part. The testing and assembly parts of the process involve probe measurements and testing, assembly and packaging, and final testing as shown in Fig. 1-3. In the wafer fabrication part, the control parameters, layout parameters, and the disturbances are represented by the C, L, and D, respectively. In-line measurements are denoted by X.

Probe and final tests are performed by testing programs that set appropriate values of the voltage and current source inputs and then measure various voltage or current outputs. Two types of tests are made: parametric and functional. Parametric tests are used to detect basic discrepancies between the performance of the IC under test and the desired performance and involve such quantities as power, critical signal levels and other DC parameters. Functional tests are used to detect errors in the function performed by IC chips. For digital circuits, functional tests involve the

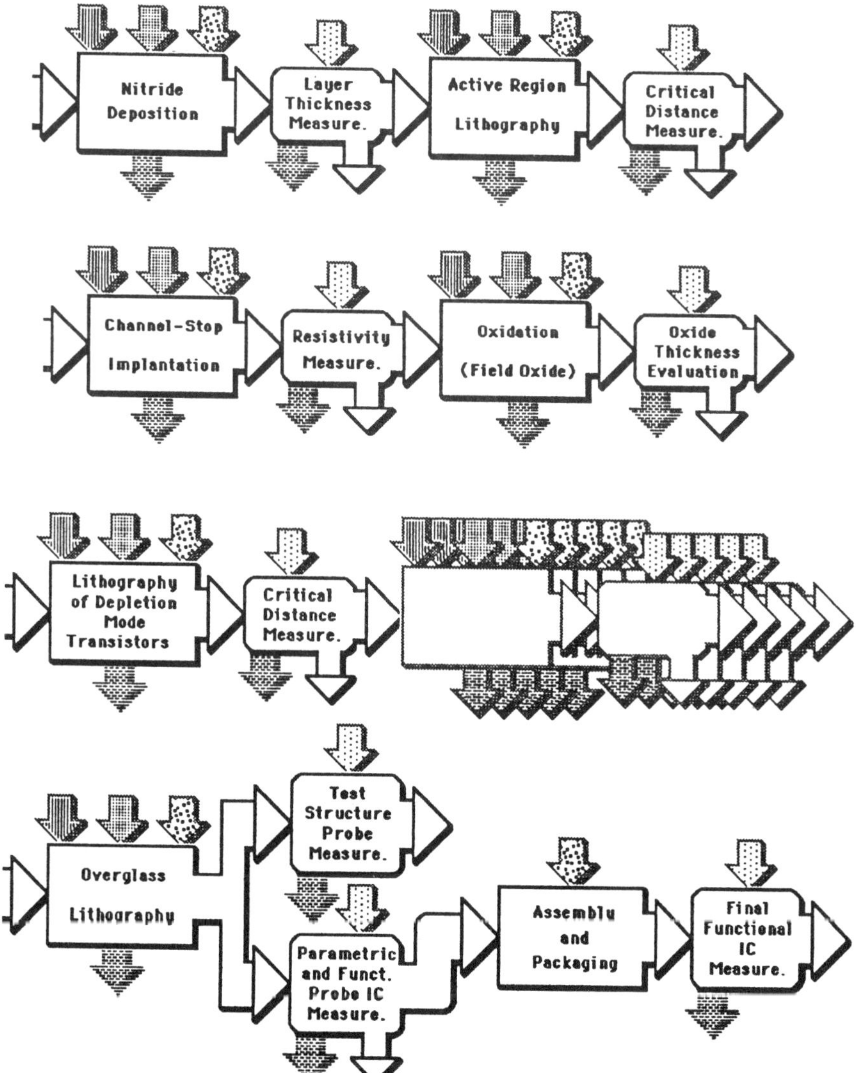

Figure 1-2: Flow of information in an NMOS process.

application of binary testing sequences to chip inputs and the results are compared against expected output binary vectors. For analog circuits, functional evaluation depends on the type of chip, but usually consists of some frequency or dynamic response measurements. A bandwidth limitation associated with the probe capacitances may be an important obstacle in the

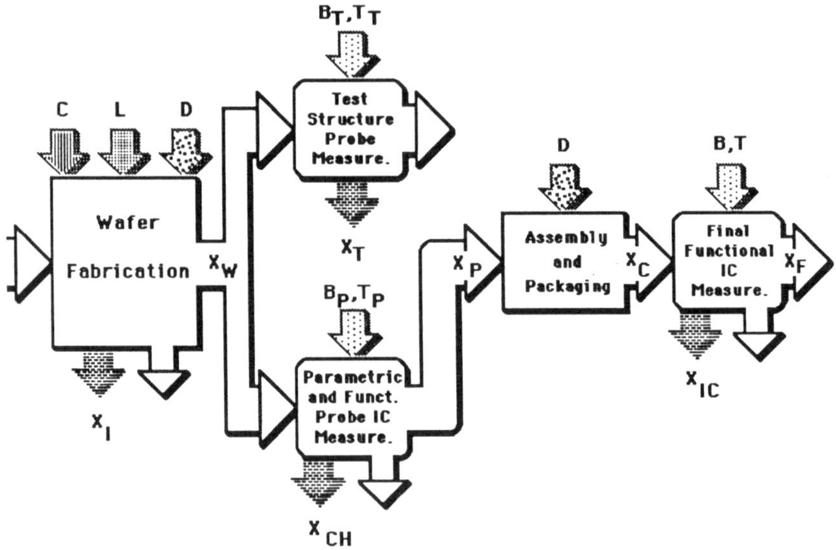

Figure 1-3: Generalized structure of the IC manufacturing process.

implementation of some functional tests during probe measurements.

We assume that the IC probe selection is characterized by two sets of parameters, B_P and T_P . B_P is a set of parametric testing conditions and binary test vectors. T_P is a set of selection thresholds which define a *selection region* . The IC probe measurement results are denoted by X_{CH} . We denote by B_T and X_T the test structure probe testing conditions and measurement results, respectively.

After probe measurement, all IC chips classified as defective are rejected and the remaining chips are assembled and packaged. A full final testing of the packaged IC is then performed, and again defective IC's are rejected. The final test conditions, selection thresholds, and measurement results are denoted by B_{IC}, T_{IC} and X_{IC}, respectively. Examples that illustrate selection thresholds for various selection tests are shown in Fig. 1-4.

In order to fully characterize the general structure of the manufacturing process, we introduce a set of variables, called *process state variables* , which completely describe the physical properties of the IC at any point in the fabrication process. While the process state variables are usually

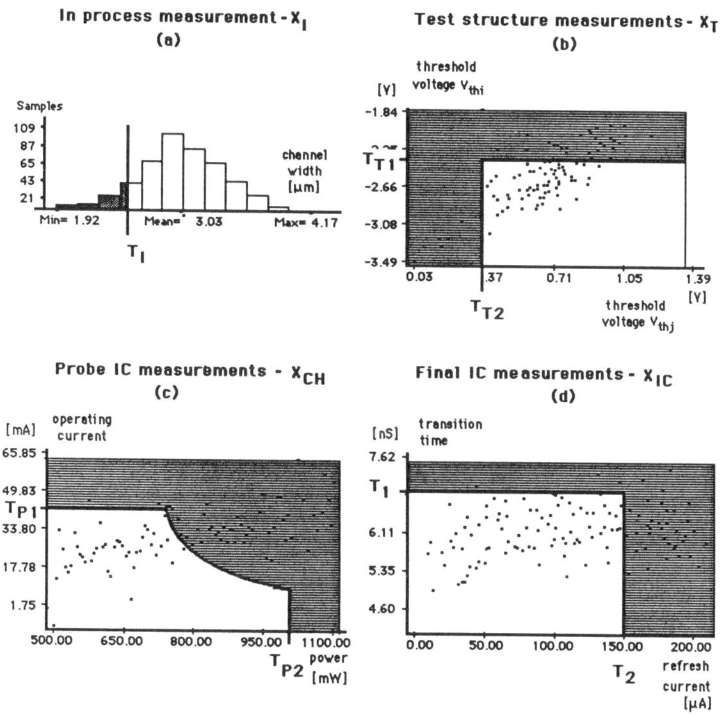

Figure 1-4: Examples of the selection regions defined for: in-process measurement, shown in the histogram (a), test structure measurements, probe measurements, and final measurements shown in the scatter plots (b), (c), and (d), respectively. IC chips or wafers whose parameters are outside of the selection region, i.e., in the shaded regions, are classified as defective and are rejected from the process.

unknown in an actual process, they can be estimated through in-line, test structure, and final measurements. We distinguish between four sets of state variables: X_W, X_P, X_C, and X_F, which describe the "state" of an IC chip after wafer fabrication, probe test, assembly and packaging, and final test, respectively, as illustrated in Fig. 1-3.

A number of relationships and dependencies exist between the variables introduced above. We define the n_i - dimensional vector X_i, $X_i = \{x_1^i, \ldots, x_{n_i}^i\}^T$, where the index i denotes the type of measurement

($i = I$ for in-line measurement, $i = CH$ for probe measurement etc.) , and T denotes transpose. X_W and X_I which describe the IC after wafer fabrication and in-line measurements, respectively, are implicit functions of the process controls, C , the layout parameters, L , and the disturbances, D , i.e.,

$$X_W = F_W(C,L,D) \tag{1.1}$$

$$X_I = F_I(C,L,D). \tag{1.2}$$

Note that F_W and F_I do not imply that explicit analytic expressions exist. Since the process disturbances, D, are random variables, X_W and X_I are random vectors and are characterized by **j**oint **p**robability **d**ensity **f**unctions (**jpdf**), which will be denoted by $f_W(x_W)$ and $f_I(x_I)$, respectively.

Similarly, we can identify the following dependencies:

$$X_T = F_T\{X_W(C,L,D),B_T\} \tag{1.3}$$

$$X_{CH} = F_{CH}\{X_W(C,L,D),B_P\} \tag{1.4}$$

$$X_P = F_P\{X_W(C,L,D),B_P,T_P\} \tag{1.5}$$

$$X_C = F_C\{X_W(C,L,D),B_P,T_P\} \tag{1.6}$$

$$X_{IC} = F_{IC}\{X_W(C,L,D),B_P,T_P,B\} \tag{1.7}$$

$$X_F = F_F\{X_W(C,L,D),B_P,T_P,B,T\} \tag{1.8}$$

where the functions $F_T, F_{CH}, F_P, F_C, F_{IC},$ and F_F are complicated relations, that usually can only be approximated from experimental data. Of course, as in the case of X_W and X_I, all of the variables on the left hand side of Eq. (1.1)-Eq. (1.8) are random vectors that are characterized by appropriate jpdf's.

To illustrate the physical significance of the above relations, consider a simple n-channel polysilicon-gate MOS technology. For this technology, we can identify the following variables:

- In-line measurements , X_I - includes field oxide thickness, gate oxide thickness, distance between edges of the field oxide in the channel of the depletion mode transistor, resistance of the polysilicon path, etc.;

- Process state vector after wafer fabrication, X_W - includes the three dimensional description of the geometry of all regions in the silicon substrate, the insulating and conducting layers, including distributions of the impurities in the entire IC chip, the location and size of all point defects in the semiconductor substrate, the parameters determining the dependence between impurity concentration and carrier mobilities, etc.;

- Test structure test conditions , B_T - includes the drain-source, substrate and gate voltages used to measure I-V characteristics of a set of transistors, the currents of the current source used to test connectivity of a chain of contacts, etc.;

- Test structure measurements X_T - includes the voltage drop across the polysilicon resistor connected to the output of the current source, the transistor drain currents measured with the gate connected to the drain for a given supply voltage for various back-bias voltages, etc. ;

- Test structure thresholds , T_T - includes the maximal value of the voltage drop on terminals of the contact chain; maximal drain substrate current, maximal and minimal voltage drops between polysilicon resistor terminals, etc. ;

- Test conditions of the IC probe measurements, B_P - includes the voltage supply, clock frequency, current at a given output node, patterns (test vectors) of 0's and 1's on IC inputs, etc.;

- Result of IC probe measurements , X_{CH} - includes input current, power dissipation, voltage gain, noise margin, offset voltage, etc.;

- IC selection thresholds , T_P - includes maximal and minimal currents of voltage supply, minimal output voltage for a given output current, patterns of 0's and 1's on the specified IC outputs, minimal amplitude of the ac signal, etc.;

- Process state after IC selection , X_P - includes parameters denoted by X_W except for those parameters that describe defective IC's. X_P could include all of the parameters that characterize IC element models provided that the element models perfectly describe the IC elements. (Examples of such parameters are the SPICE MOS transistor model parameters.);

- Process state after assembly and packaging , X_C - includes parameters by X_P as well as the data that characterize packaging, i.e. the data that describes power dissipation conditions etc.;

- Results of final functional test conditions and test sequences, B - includes parameters denoted by B_T as well as a full set of test vectors and all high frequency and high speed conditions such as high frequency clock rates, high frequency ac signals etc.;

- Final functional test and measurements , X_{IC} - includes those parameters denoted by X_{CH} as well as the full set of results of functional test and high frequency measurements such as frequency characteristics and delays;

- Final functional test thresholds , T - includes the complete set of the correct results of functional tests and quantities such as maximal delay, minimal amplitude of the ac signal, and all other thresholds included in T_P;

- Process state after final testing , X_F - is equivalent to X_C but can also be expressed in terms of a set of parameters of IC device models that fully characterize IC behavior under all possible test conditions.

In conclusion, we emphasize the following facts:

- Due to process disturbances, all quantities observed during manufacture are random variables and must be described by appropriate jpdf's;

- Process state variables are essentially unknown and can only be estimated by means of measurements;

- Relationships between process conditions and the quantities measurable in the process are usually known with limited accuracy and in many cases is experimental in nature.

We now investigate the causes of process fluctuations, and their effects on circuit performance.

1.3. Disturbances in the IC Manufacturing Process

The process disturbances that occur in the IC manufacturing process can be characterized in terms of the physical nature of the disturbance or in terms of the effects they have on IC performance. Effects on performance can be described in terms of changes in the electrical characteristics of the IC or in terms of changes in circuit connectivity. In this section we describe and classify process disturbances and then classify the effects that these disturbances have on performances.

1.3.1. Process Disturbances

Sources of the random phenomena that occur in the IC fabrication process [118], [34], [96] which can be modeled as process disturbances, can be classified as:

- **Human errors and equipment failures**.

- **Instabilities in the process conditions**. These are random fluctuations in the actual environment that surrounds an IC chip and arise from phenomena such as turbulent flow of gasses used for diffusion and oxidation, inaccuracies in the control of furnace temperature, etc. Because of these instabilities, each area of wafer is exposed to slightly different environmental conditions and, hence, no two manufactured chips can possibly have identical performance.

- **Material instabilities**. These are variations in the physical parameters of the chemical compounds and other materials used in the manufacturing process. Typical examples of material instabilities are: fluctuations in the purity and physical characteristics of the chemical compounds, density and viscosity of photoresist, water and gasses contamination etc.

- **Substrate inhomogeneities**. These are local disturbances in the properties of substrate wafers and are of three types: *point defects*, *dislocations*, and *surface imperfections*. Point defects, which are local disorders in the structure of the lattice of a semiconductor material include: *vacancies* (a lack of the atom in the lattice site), *interstitial defects* (an atom is located out of crystal site), and *vacancy-interstitial pairs* (also called *Frenkel defects*). Point defects act as recombination centers and, therefore, affect carrier lifetime. Concentration of point defects may fluctuate and is affected by high temperature process operations. Dislocations, which are geometrical irregularities in the regular structure of the crystal lattice, are caused by crystal growth related thermal stresses and are of two different types: *edge dislocations* and *screw dislocations*. The most important characteristic of a dislocation is that it strongly interacts with point defects and impurities located in its neighborhood. As a consequence, a large increase of the impurity diffusion coefficient may occur which may lead to the formation of diffusion pipes that cause such problems as abnormal leakage currents or low breakdown voltage in p-n junctions.

- **Lithography spots**. These are lithography related disturbances of IC geometry which may be caused by the mask fabrication process as well as the lithography process itself. The mask fabrication process may produce defects of two types: transparent spots in opaque regions and opaque spots in transparent regions of the mask. The lithography process may produce spots due to dust particles on the surface of the mask and photoresist layers. Some photoresist inhomogeneities can also produce the lithography spots.

Not all of the disturbances described above have the same influence on IC performance. Thus, we are motivated to consider what electrical and geometrical effects result from these disturbances.

1.3.2. Process Related Deformations of IC Design

Disturbances in the process can be viewed as affecting either the geometry and/or the electrical characteristics of an IC device. In order to characterize these effects it is useful to employ the concept of an *ideal IC*. An ideal IC is one that would result from a perfect manufacturing process, ie., its layout is identical to that specified by the masks, and its electrical properties (e.g. impurity distributions or charge densities) are the same as those assumed by the designer during the design process. The performance of an actual IC can be viewed as a deformation from that found in an ideal IC. Actually, deformations have two components: a deterministic component, and a random component. The deterministic component of each deformation is due to simplifications used in developing the description of the ideal IC. Such simplifications are usually necessary due to the lack of the adequate process/device modeling techniques applied during the IC design process. The random component of the deformation is due to the process disturbances themselves. We are particularly interested in this random component.

1.3.2.1. Geometrical Deformations

The geometry of an IC is determined by the two dimensional (2D) IC layout and the physics of the manufacturing process. In an ideal IC it is assumed that the edges of the layout define the boundaries of the various regions in an IC and that various manufacturing process steps determine the depth, or thickness, of a region. In an actual IC this is not the case, as illustrated in Fig. 1-5. Geometrical deformations result from three effects: *lateral effects*, *vertical effects*, and *spot defects*.

Lateral effects include all processing effects that cause the location of the boundary of the region in an actual IC to differ from the corresponding boundary in the ideal IC. A lateral effect can be composed of a *lateral edge displacement* and a *mask misalignment*.

Lateral edge displacements are caused by all processing steps with anisotropic characteristics, such as:

- Lateral diffusion;

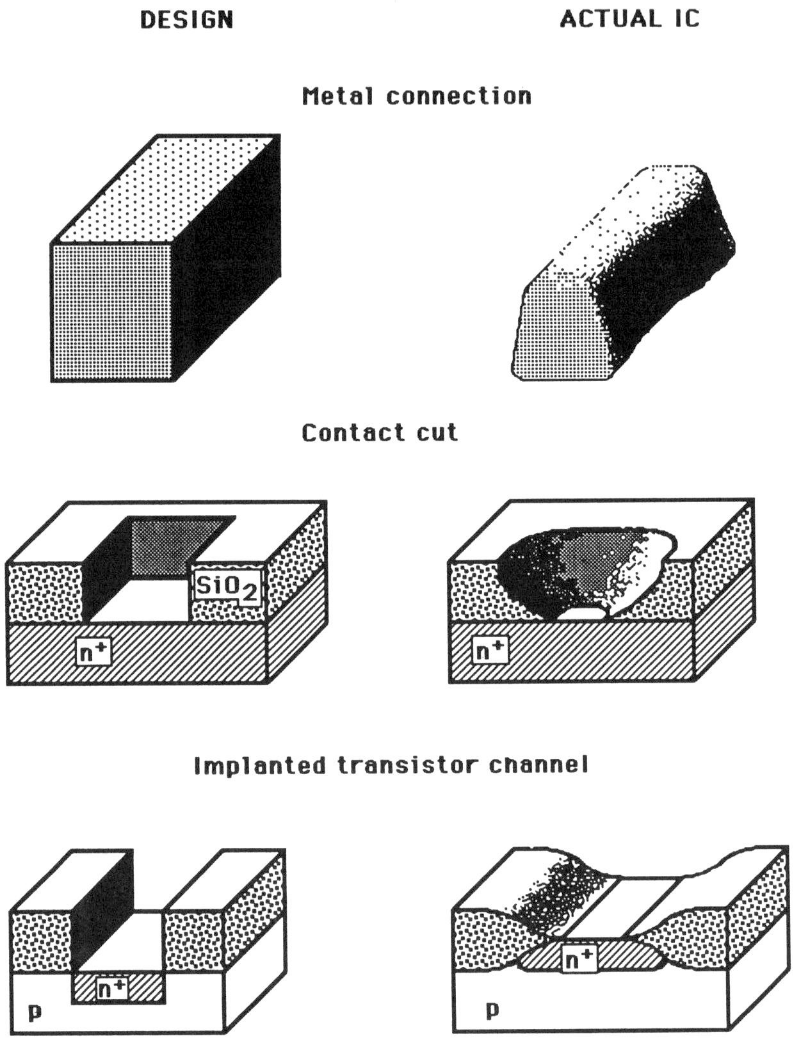

Figure 1-5: Illustrations of the discrepancies in geometry that exist between an ideal IC and an actual IC.

- Lateral oxidation i.e., bird's beak formation;
- Over- and under-etching of metal, polysilicon, oxide and nitride layers;

- Over- and under-exposure of the photoresist and photoemulsions apllied in all lithography steps;

- UV light or e-beam diffraction on the mask edges.

Fig. 1-6 illustrates the contribution of various lateral effects to the lateral edge displacement of the diffusion region, bounded by the edge of the field oxide. Lateral edge displacements may cause drastic deformations in the topology of the actual IC by merging regions which are supposed to be disjoint. In terms of electrical behavior such deformations manifest themselves as shorts between, or breaks in, conducting paths.

Note that all edges that lie within regions of the same type are usually affected by the same set of independent disturbances, while edges that lie in different types of regions are affected by different sets of disturbances. There are cases, however, in which some disturbances affect different types of regions in the same way and where regions of the same type are affected by quite different sets of disturbances. For example, in n-channel MOS technology, some of the disturbances that contribute to the displacement of the diffusion edge, as depicted in Fig. 1-6, affect both the location of the edge of the MOS transistor drain and source as well as the location of the edges of transistor channels. On the other hand, the locations of the edges of depletion mode transistors are determined by two different sets of disturbances. Specifically, the edges on the boundaries between channel and source or channel and drain are affected by different disturbances than the edges of the active regions. Hence, in general, in an actual IC, displacements of the edges of a region may either be very similar (if not identical), correlated, or totally independent.

Disturbances that significantly contribute to lateral edge displacements are random instabilities in process conditions such as equipment vibrations, and instabilities of layer thicknesses that cause fluctuations in the amount of over- and under-etching.

Mask misalignment , which is an error in the position of a lithography mask with respect to the features already engraved on the surface of a wafer, cause all edges defined by the mask to be shifted by the same amount with respect to the boundaries of the regions that already exist on the surface of the wafer. (See Fig. 1-7) In fact, mask misalignment could be so significant that the geometry of an actual IC can be deformed from that of the corresponding ideal IC to the extent it results in changes in electrical

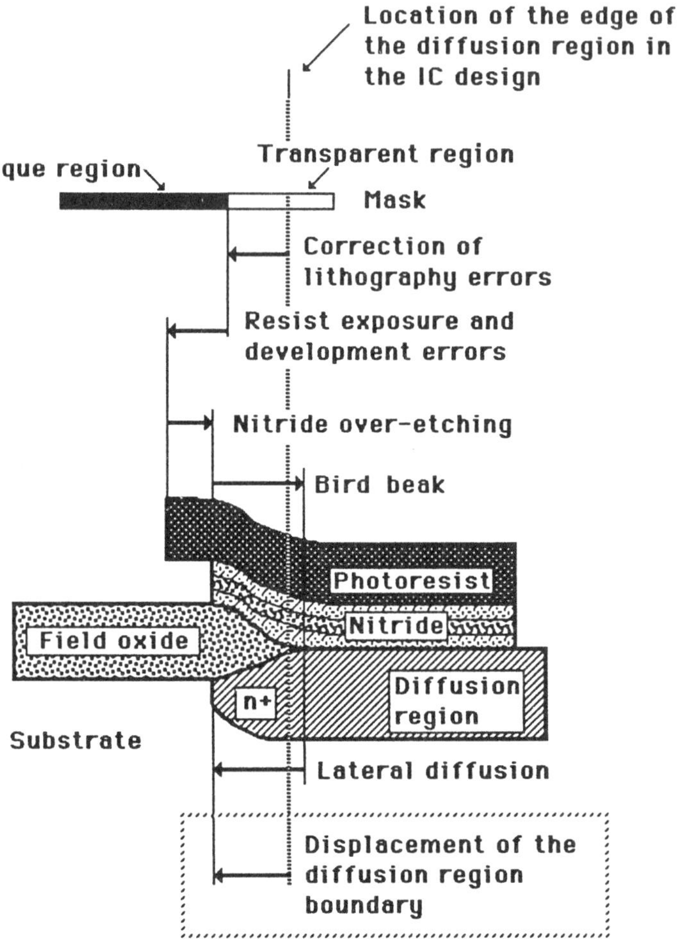

Figure 1-6: Lateral displacement of the diffusion region boundary as a result of edge displacements in the mask fabrication process, resist and nitride etching, bird's beak formation and lateral diffusion.

connectivity. Random errors in the mask alignment process are due to instabilities in process conditions due to limited mechanical and optical accuracy of the processing equipment and shape variations of the wafers.

Vertical effects are deformations in the thickness of IC layers and include

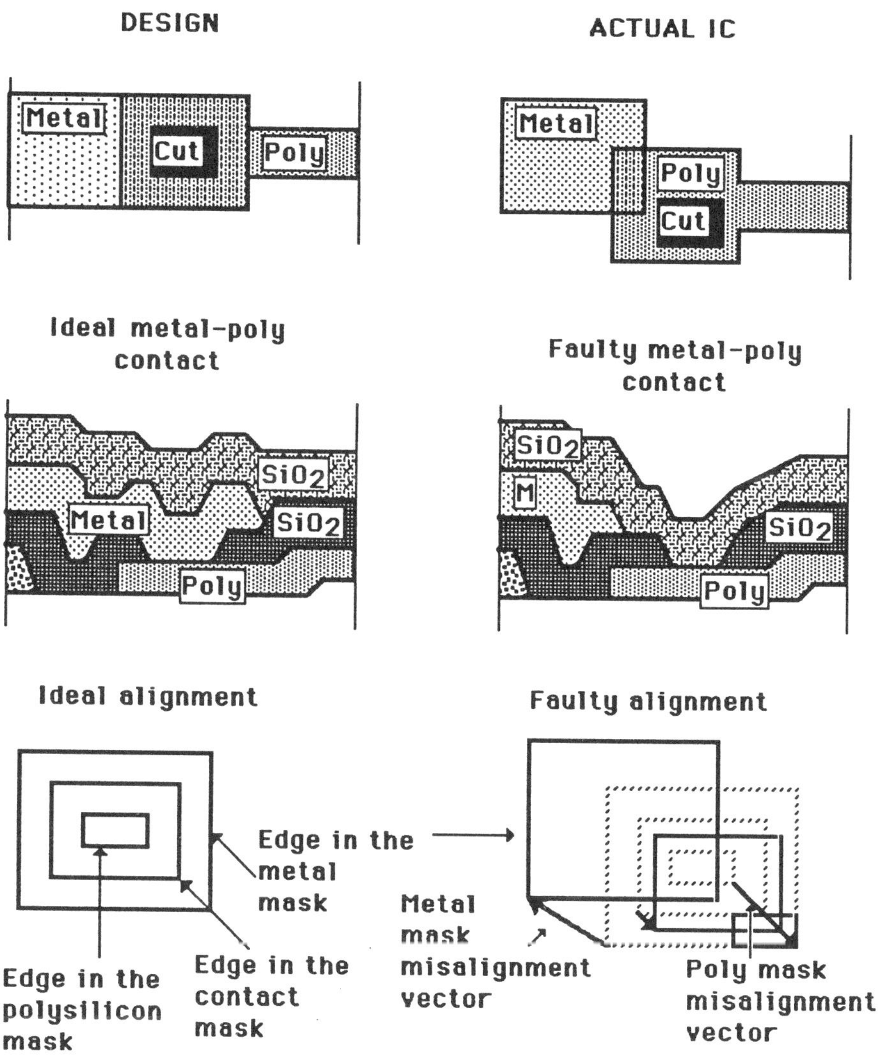

Figure 1-7: Mask misalignment.

deformations which are due to the p-n junction depth variations and deformations in the thickness of the oxide and other deposited layers. Junction depth variations are a direct consequence of the fluctuations in the impurity concentrations while deformations in the thickness of the deposited or oxidized layers are due to process instabilities such as turbulent gas flow, temperature fluctuations, etc.

Spot defects are geometrical features that emerge during the manufacturing process that were not originally defined by the the IC layout. The main source of spot defects are lithography spots. Spot defects are random in both their location and size. The size of a spot defect is determined by both the size of the actual particle that caused the spot defect and lateral deformations which disturb the location of the edges of the spot defect.

1.3.2.2. Electrical Deformations

The electrical characteristics of an actual IC are determined by the three-dimensional impurity distributions in the conducting and semiconducting regions, and charge distribution in the insulating layers. In an ideal IC, a number of simplifying assumptions are typically made which allow for analysis of a design prior to manufacturing (see Fig. 1-8). All process models are described by simplified equations that can be solved either analytically or by simple numerical techniques. Thus, the physical description of such phenomena as dependence of impurity diffusivities on concentration of vacancies and interstitials, or dependence of oxidation growth rates on the impurity concentration in the substrate is often inadequate and leads to significant discrepancies between simulation results and reality. Furthermore, most models are one-dimensional and incorporate only a few of two-dimensional effects. Device analysis is also performed in an approximate manner and although the equations that describe device behavior are known, the boundary conditions are typically simplified. Also , a number of important physical effects are not included in the analysis (for example, heavy doping effects in bipolar transistors or thin oxide tunneling effects in MOSFET's).

The above simplifications introduce a systematic error in performance analysis and cause the description of the electrical properties of an IC used in the design process to differ from physical reality. However, there also exist many phenomena that are random in nature and cause electrical deformations of IC characteristics. As was the case for geometrical deformations, we can divide these phenomena into global and local phenomena. Global electrical deformations affect various regions in different IC elements in the same way. Examples of global deformations include:

- fixed positive oxide charges in the oxide due to a non-perfect (non-stechiometric) structure of the oxide;

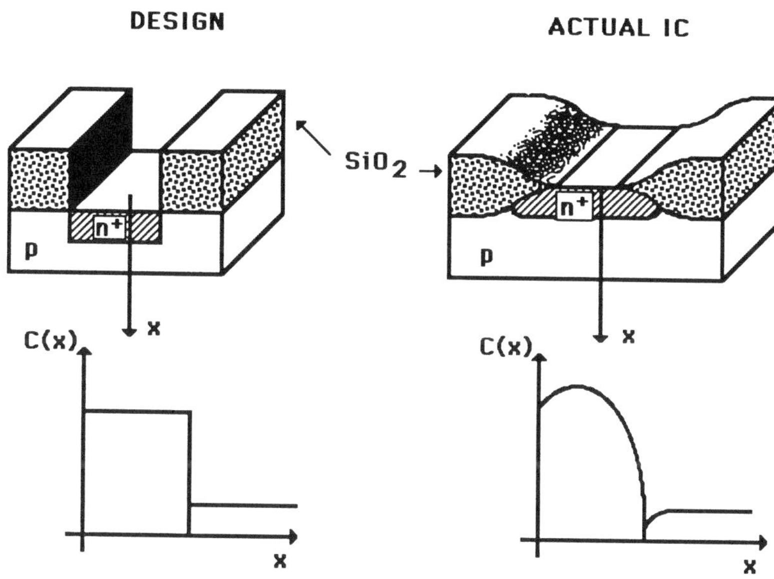

Figure 1-8: Comparison of impurity distributions assumed in the channel of an ideal depletion mode transistor versus that of an actual depletion mode transistor.

- emitter-push effect, which is due to enlarged diffusivity of boron in the active base of the n-p-n transistor, caused by emitter diffusion induced damages of the crystal lattice;

- increased junction depths under the oxide regions due to oxidation-enhanced diffusion phenomena.

These deformations are random and vary from wafer to wafer, and even across the wafer. However, within one chip their effects on IC elements are very similar. Local electrical deformations are due to

- Spikes, which are tiny conducting pipes, or paths, crossing junctions or even whole regions, e.g., the base in the n-p-n transistors;

- Local impurity precipitates or silicides which affect the percentage of electrically active impurities in silicon;

- Local contaminations of the SiO_2 - Si surface that affect surface recombination mechanisms in bipolar devices and disturb MOS transistor threshold through local concentration of surface states.

The above effects have to be treated as process related deformations of the electrical characteristics of an actual IC and have to be taken into account in order to obtain a realistic model for yield losses.

1.3.3. General Characteristics of Process Disturbances

From the above discussion we can observe that disturbances can cause either global or local deformations. All deformations which affect all IC elements in the same way are called *global*. Mask misalignment is an example of a disturbance that causes a global deformation of an ideal IC. Disturbances that cause global deformations are referred to as global disturbances . Similarly, disturbances that cause only local deformations (affected regions are small compared to the total area of an IC chip) are referred to as local disturbances . Pin-holes, spikes and spot defects are examples of local deformations.

It is important to note that various process conditions may interact in an indirect way. For instance, high temperature processes may cause an increase in the lithography errors due to deformations in the shape of the wafer. Hence, in general, random fluctuations in process conditions of one processing step may result in changing process conditions of another step. Note also that process induced deformations of any kind, local or global, geometrical or electrical, are random. For the case of global disturbances, the magnitude of process-related deformations is random and varies from one IC to another. For the case of local disturbances, both the magnitude (e.g. size of a defect) and location of the resulting deformations are random.

1.3.4. IC Performance Faults

We now consider the impact that process disturbances have on IC device characteristics and indicate how these randomly changing characteristics affect IC performance. We then introduce a fault classification based upon effects on performance.

We begin by noting that each process disturbance may cause a number of different changes in IC performance. Such changes in behavior are called *faults* . For instance, a spot defect introduced in a lithography step may cause a short between two adjacent conducting paths, a break in the path or may only alter effective device dimensions (e.g. affect the W/L ratio of a MOSFET). Faults may be classified as either ***structural faults***, which are changes in the topology of the circuit or electrical diagram, that represents an IC, or *performance faults* , in which the IC topology remains unchanged, however, circuit performance (e.g., speed or dissipated power) falls outside of some set of allowable tolerance limits. Performance faults can be further divided into two classes: *soft-performance faults* and *hard-performance faults* . If an IC is functionally correct (e.g. in a digital circuit, all state transitions occur in the proper order and the output pattern is the correct response to a given input pattern vector) but some performance measure (e.g. signal delay) lies outside of the specified range, we say a soft-performance fault has occurred. If an IC does not function properly (e.g. some state transitions do not occur) or some performances are orders of magnitude from the desired values, we say a hard-performance fault has occurred.

Structural faults require some further elaboration. While some deformations will not alter the structure of an IC under no bias, or very small bias conditions, for other values of DC bias a short between two conducting paths can occur. As an example, consider the lateral diffusion in two parallel interconnect lines. Due to the expansion of depletion layers of the p-n junctions, there might be an equivalent short between these two lines under some bias conditions. On the other hand, some geometrical variations may introduce new parasitic elements (such as overlap capacitances) which do not necessarily cause an IC to be defective. Therefore, structural faults are best defined in terms of changes in the equivalent DC circuit diagram under worst case bias conditions.

This fault classification scheme is illustrated in Fig. 1-9. The bottom of the figure shows our classification of physical phenomena which cause yield loss while in the upper part of the figure, classification is based upon effects (performance faults, structural faults). Note that each phenomenon can be a cause for a variety of faults. However, some of the relations (solid lines) depicted in Fig. 1-9 dominate. For instance, all global effects are primarily responsible for all soft performance failures while point defects are primarily responsible for structural failures.

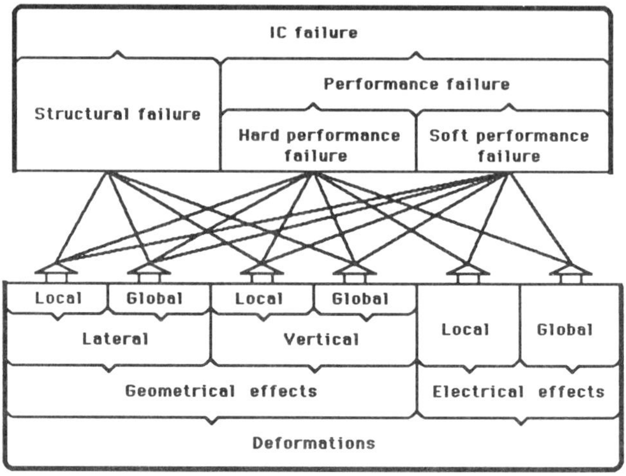

Figure 1-9: Classification of IC failures and process deformations.

1.4. Measures of Process Efficiency

As discussed above, the inherent fluctuations in an IC fabrication processed above, that cause variations in circuit performance so that not all of the chips that are manufactured actually meet all of the design specifications. We define the *manufacturing yield* as the ratio of the number of chips that successfully pass all of the selection steps in the process to the total number of chips that begin the fabrication process. Thus manufacturing yield is a measure of process efficiency . Since manufacturing yield is affected by both the parameters of the design (e.g. layout) as well as the parameters of the manufacturing operations themselves (e.g. control and disturbances), it is desirable to be able to *predict yield* during the design process, as well as *estimate yield* during the manufacturing process. By being able to predict yield during the design process, it may be possible to adjust the design to improve yield. Similarly during manufacturing, if the estimated yield is low, changes in the process which increase yield may be possible. Thus we are motivated to develop techniques for yield estimation and prediction.

1.4.1. Yield Estimation

Before we begin our discussion of yield estimation , it is convenient to introduce the following notation (see Fig. 1-10) :

- N is the maximum number of IC chips that can possibly be fabricated on W production wafers, assuming an ideal manufacturing process;
- W_W is the number of wafers that have reached probe selection tests;
- N_W is the number of IC chips on W_W wafers;
- N_P is the number of chips that have reached final selection step;
- N_F is the number of chips that have been classified as fault-free after the final selection step;
- N_I is the number of chips that have been rejected from the process during the in-line selection step;
- N_{CH} is the number of chips that have been rejected from the process in the probe selection;
- N_{IC} is the number of chips that have been rejected during final selection;
- N_S is the number of test structures measured in the probe measurements;
- N_T is the number of test structures classified as correctly fabricated.

We can now formally define a variety of yield estimates. As will be discussed later, each of these estimates are useful in assessing the efficiency of different parts of the manufacturing process.

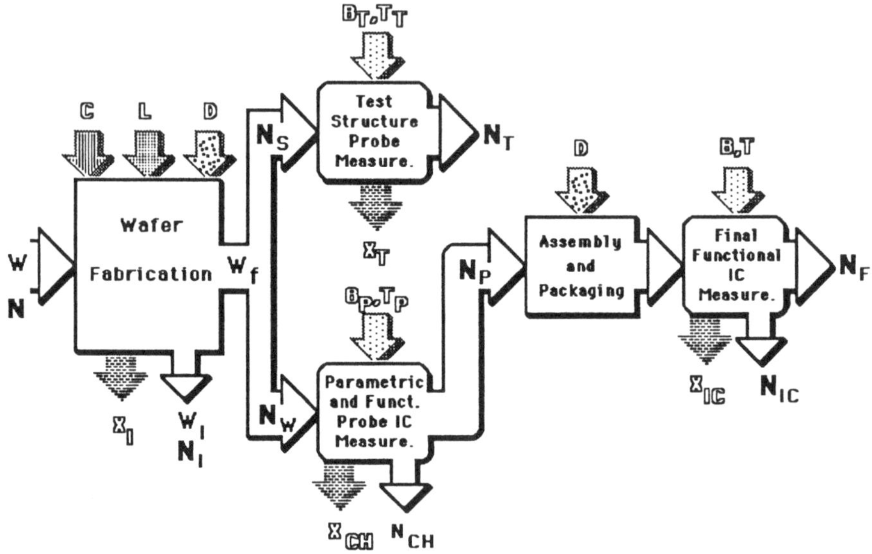

Figure 1-10: Flow of IC chips in the manufacturing process.

1. The *manufacturing yield* :

$$Y_M^* = \frac{N_F}{N} \tag{1.9}$$

is a measure of an overall process efficiency. It indicates overall manufacturing performance and is a basic figure in evaluating manufacturing economics and planning manufacturing strategy.

2. The *probe yield* :

$$Y_P^* = \frac{N_P}{N} \tag{1.10}$$

describes the efficiency of the wafer processing operations including probe measurements.

3. The *wafer yield* :

$$Y_W^* = \frac{W_W}{W} \tag{1.11}$$

which can also be expressed as:

$$Y_W^* = \frac{N_W}{N} \tag{1.12}$$

is a measure of the quality of the wafer fabrication operations. It is the percentage of manufacturing wafers that have reached probe test. In practice, the majority of wafer rejections is due to human errors or equipment failures. Therefore, wafer yield is primarily a measure of the quality of process supervision.

4. The *final testing yield* :

$$Y_{FT}^* = \frac{N_F}{N_P} \tag{1.13}$$

is the percentage of IC chips that passed final test. Note that some IC chips are rejected from the process due to assembly and packaging errors and functional defects that were not detected during probe selection.

5. The *probe testing yield* :

$$Y_{PT}^* = \frac{N_P}{N_W} \tag{1.14}$$

is a measure of wafer fabrication efficiency. For the case of new products, probe testing yield shows whether a fabricated IC was designed to match the manufacturing line capabilities. For correct designs, the probe testing yield is the best indicator of the process instabilities.

6. The *processing yield* :

$$Y^*_{PRO} = \frac{N_T}{N_S} \tag{1.15}$$

is a measure of the discrepancy between the performance of the process and desired process parameters described in terms of model parameters of the IC elements. The processing yield is used to determine manufacturing line stability.

It should be stressed that $N_W, N_P, N_F, N_{CH}, N_{IC}, N_S$, and N_T are random and, of course, have jpdf's determined by the relations characterized by Eqs. Eq. (1.1) to Eq. (1.8) and therefore are related to $D, L,$ and C as well as to the testing conditions. Thus by careful analysis of the values of yields defined above one can the perform process diagnosis, which can be viewed as estimating D, as well as evaluating the quality of process-design tuning which can lead to the corrections in L and C .

1.4.2. Yield Prediction

Note that evaluation of each of the yield estimates defined above requires the use of actual process measurements and is, therefore, inadequate for predicting yield during the design process prior to fabrication. In order to predict yield during the design process we need to be able to determine the probability that a single IC chip will not be rejected during fabrication. To evaluate this probability it is convenient to introduce the concept of an *acceptability region* .

Let the n_F dimensional vector $Q_F = \{ q^F_1, \dots, q^F_{n_F} \}^T$ denote all of the functional performance measures of interest for a particular design, and the n_P dimensional vector $Q_P = \{ q^P_1, \dots, q^P_{n_P} \}^T$ denote all of the parametric performance measures of interest for the same design. For a digital IC, functional performance is a measure of the ability to perform some desired logical operations. Parametric performance is a measure of the quality of the behavior of an IC. Quantities such as delay, power, input and output current, gains, etc. are typical measures of parametric performance. Let the n_Q dimensional vector $Q = \{Q_F, Q_P\} = \{ q_1, \dots, q_Q\}$ ($n_Q = n_F + n_P$) denote all performances of interest. We now define the n_F dimensional space $\mathbf{S}_Q^F$ and n_P dimensional space $\mathbf{S}_Q^P$ to be Euclidean spaces that contain all n_F functional performances and n_P parametric performances, respectively. Finally, we define the n_Q dimensional *performance space* as $\mathbf{S}_Q = \mathbf{S}_Q^F \times \mathbf{S}_Q^P$. We can now define the set of *acceptable IC performances* as

$$\mathbf{A}_Q = \{Q \in \mathbf{S}_Q \mid \text{each } q_j \text{ isacceptable} \ \forall \ j = 1,2,\ldots,n_Q\} \tag{1.16}$$

Since the performance of an IC is fully determined by either the vector of process state variables X_W, which represent all of the physical parameters of the IC after wafer processing, or the variables L, C , and D, which actually define the design, the set of acceptable performances $\mathbf{A}_Q$ must have its equivalent in the n_W dimensional space of process state variables $\mathbf{S}_W$. Hence, we can define the *acceptability region*

$$\mathbf{A}_W = \{X_W \in \mathbf{S}_W | Q(X_W) \in \mathbf{A}_Q\} \tag{1.17}$$

where $Q(X_W)$ symbolizes the dependence of the performance on the process state variables. In other words, the acceptability region is defined by a set of process state variables that assure the desired performance of the fabricated IC. Note that this region, unlike the selection regions, is not dependent on the method by which IC measurements, or measurement conditions, are defined.

Now using the concept of the acceptability region we can define *design yield* , Y, as the probability that the IC performs all functions correctly. Such a probability can be expressed as:

$$Y = \int g(x_W)\, f_W(x_W)\, dx_W \tag{1.18}$$

where $f_W(x_W)$ is jpdf associated with the random variable X_W and

$$g(x_W) = \begin{cases} 1, & \forall \ x_W \in \mathbf{A}_W \\ 0, & \text{otherwise} \end{cases} \tag{1.19}$$

Equations Eq. (1.18) and Eq. (1.19) can also be expressed as the integral:

$$Y = \int_{\mathbf{A}_W} f_W(x_W)\, dx_W \tag{1.20}$$

where the integration is taken over all values of $X_W \epsilon \mathbf{A}_W$.

Design yield can also be expressed in the process disturbances space $\mathbf{S_D}$, which is an Euclidean space that contains all of the process disturbance given in the vector D . We define the acceptability region in space $\mathbf{S_D}$ as follows. For specific values of the variables C and L , denoted by C^o and L^o, a disturbance vector δ will be in the acceptability region, that is denoted by $\mathbf{A}_D(C^o,L^o)$, if $F_W(C^o, L^o, \delta) \epsilon \mathbf{A}_W$ where F_W is defined by Eq. Eq. (1.1). In other words, there exists, for a given L^o and C^o , in the space of process disturbances a set of disturbances $\mathbf{A}_D(C^o,L^o)$, which will not cause the performance of an IC to be unacceptable. This set can also be viewed as the projection of the acceptability region $\mathbf{A}_W$ onto the space $\mathbf{S^D}$. Hence, we can express design yield as

$$Y = \int_{\mathbf{A}_D(C^o, L^o)} f_D(\delta)\, d\delta \tag{1.21}$$

where $f_D(\delta)$ is the jpdf which describes D.

The design yield , as defined in Eq. (1.18) and Eq. (1.19) or Eq. (1.21) is difficult to evaluate, because in general it is not possible to derive an analytic expression that relates IC performance to the disturbance, layout and control parameters. In other words it is not possible to obtain an explicit expression for $\mathbf{A}_D(C^o,L^o)$. Furthermore, some of the components of D can only be vaguely defined and only a subset of state variables are actually known. However, through an analysis of the basic properties of the process disturbances, as was carried out in the previous section we can develop techniques for evaluation of design yield.

1.4.3. Decomposition of the Design Yield Equations

We may view $\mathbf{A}_Q$ the acceptability region as being composed of the set of *functionally acceptable performances*

$$\mathbf{A}_Q^F = \{ Q \in \mathbf{S}_Q^F \mid \text{each } q_j^F \text{ is acceptable } \forall j = 1,2,\ldots,n_F\} \tag{1.22}$$

and the set *parametrically acceptable performances*

$$\mathbf{A}_Q^P = \{Q \in \mathbf{S}_Q^P \mid \text{each } q_j^P \text{ is acceptable } \forall j = 1,2,\ldots,n_P\} \tag{1.23}$$

Existence of two different types of performance leads us to consider two types of acceptability regions: one which assures functional correctness of the IC behavior, and another one, which assures acceptable quality of performance. Specifically, in the space of process state variables $\mathbf{S}_W$, we can define the *functional acceptability region*,

$$\mathbf{A}_W^F = \{X_W \in \mathbf{S}_W \mid Q(X_W) \in \mathbf{A}_Q^F \} \tag{1.24}$$

and the *parametric acceptability region*

$$\mathbf{A}_W^P = \{X_W \in \mathbf{S}_W \mid Q(X_W) \in \mathbf{A}_Q^P \}. \tag{1.25}$$

Consequently, in the space of process disturbances $\mathbf{S}_D$, assuming a given L^o and C^o, we can define equivalents of functional and parametric acceptability regions as sets of disturbances which do not cause the IC to have incorrect functionality or poor quality. We denote these sets by $\mathbf{A}_D^F(C^o,L^o)$ and $\mathbf{A}_D^P(C^o,L^o)$, respectively. Thus, we can define the *functional yield* as:

$$Y_{FUN} = \int_{\mathbf{A}_D^F(C^o,L^o)} f_D(\delta)\, d\delta \tag{1.26}$$

where as before $f_D(\delta)$ is the jpdf describing D, and the *parametric yield* as:

$$Y_{PAR} = \int_{\mathbf{A}_D^P(C^o, L^o)} f_D(\delta)\, d\delta \tag{1.27}$$

which are probabilities that the IC is correct functionally and that its performance characteristics are within some desired limits.

Decomposition of the design yield into parametric and functional components provides an opportunity for significant simplifications in evaluating design yield. This becomes evident if we note that, since $\mathbf{A}_D^F(C^o, L^o)$ contains those process disturbances that do not disturb functional correctness of the IC, then $(1 - Y_{FUN})$ can be estimated by searching for the complement of $\mathbf{A}_D^F(C^o, L^o)$, i.e. determining the set of process disturbances that cause structural failure and/or hard performance faults only. Thus, estimating $(1 - Y_{FUN})$ we have to take into account such defects as dislocations, spot and point defects, stacking faults, oxide pinholes, etc., which cause structural changes in the connectivity of an IC but are local in nature. As a consequence, Y_{FUN}, as defined by Eq. (1.26), may be approximated by Y'_{FUN} given by:

$$Y'_{FUN} = \int_{\mathbf{A}_D^F(C^o, L^o)} f_{D'}(\delta')\, d\delta' \tag{1.28}$$

where $f_{D'}(\delta')$ is the jpdf of D', which describes only the subset of D that characterizes local processing defects. By dealing with D', instead of D, we do not need to evaluate IC performance in order to determine $\mathbf{A}_D^F(C^o, L^o)$ because process disturbances that cause structural changes in the connectivity of an IC can be found by analyzing the layout of the IC and the statistical characterization of the local defects only. In particular to determine $\mathbf{A}_D^F(C^o, L^o)$, we need only investigate the *critical area* of a layout, i.e. that area an IC which is vulnerable to the occurrence of a point or spot defect. Hence, Y'_{FUN} can be estimated by using one of a number existing geometrical approaches [79], [106], [107], [129], [106], [87], [9], [66], [68].

Computation of parametric yield is also simplified using this decomposition due to the fact that a single spot or point defect, if it does not change the circuit connectivity, usually does not affect the quality of IC performance.

So, local defects can be discarded from consideration when computing parametric yield. Hence, $\mathbf{A}_D^P(C^o,L^o)$ can be estimated in the subspace of $\mathbf{S}_D$, denoted by $\mathbf{S}_D''$, that contains only global deformations, such as mask misalignments and lateral edge displacements. The effect that these disturbances have on IC performance can be effectively modeled [65] and [83]. Hence, parametric yield:

$$Y'_{PAR} = \int_{\mathbf{A}_D^F(C^o, L^o)} f_D''(\delta'')\, d\delta'' \tag{1.29}$$

where $f_D''(\delta'')$ is the jpdf which describes D'', can be estimated. In fact, it is exactly this type of yield that is usually used as the objective function for yield maximization [23], [90], [24], [7], [120], [91], [13], [55], [3], [116]. We discuss this further in Chapter 2.

1.5. Discussion

In Section 1.4.1, we introduced a number of yield estimates and methods for yield prediction. Experimentally-based yield estimates are useful for evaluating the efficiency of different parts of the IC fabrication process and can be used to diagnose the cause of yield losses. In this section, we discuss the relationships between these estimates in order to facilitate an understanding of the physical meaning of each estimate. We then illustrate how this information may be used to analyze a manufacturing process and relate it to production cost.

1.5.1. Relationships Between Manufacturing and Design Yields

As we indicated above, the manufacturing yield Y_M^* is a good measure of the overall efficiency of a fabrication process. Observe that manufacturing yield can be viewed as the product of three other yield estimates: wafer yield Y_W^*, probe testing yield Y_{PT}^*, and final testing yield Y_{FT}^*:

$$Y_M^* = \frac{N_F}{N} = \frac{N_W}{N}\frac{N_P}{N_W}\frac{N_F}{N_P} = Y_W^* Y_{PT}^* Y_{FT}^* \tag{1.30}$$

By evaluating and comparing the values of these three factors, we can identify the principal reason for a low yield in a manufacturing process. We will return to this point when we analyze manufacturing cost.

The manufacturing yield Y_M^* is typically close to the design yield, Y. However, in reality, IC chips cannot be tested for all possible faults (in other words, the test coverage will always be less than 100%) and therefore, Y_M^* is always an overestimation of Y:

$$Y_M^* \geq Y \tag{1.31}$$

Observe that both testing programs (probe and final) are typically arranged in such a way that the parametric tests are grouped together and functional tests are grouped together. If we estimate the yield components from these selective measurements, these estimates should be close to the estimates derived in the previous section. However, due to the fact that in an evaluation technique we strictly separated fault types, the values obtained from measurements will be less than theoretical estimates. Similarly, because this strict separation of faults does not actually occur in reality, the actual yield Y will be less than the calculated product of the parametric and functional yields:

$$Y \leq Y'_{PAR}\, Y'_{FUN} \tag{1.32}$$

Finally, if the acceptability region in terms of device parameters is properly established, the processing yield Y_{PRO} defined in terms of test structure measurements, should be approximately equal to the parametric yield Y_{PAR}:

$$Y_{PRO} \approx Y_{PAR} \tag{1.33}$$

Hence, design yields can be used to predict or derive bounds for some manufacturing yields. In general, however, design and manufacturing yields have been defined for different purposes, and therefore, the accuracy of yield prediction should not be evaluated in terms of a simple comparison with the manufacturing yield.

1.5.2. Examples of Yield Analysis

The above discussion attempts to present a unified framework for yield estimation and prediction. However, in the past, much simpler approaches to yield estimation and prediction sufficed. For example, for some types of circuits, it is well known that the manufacturing yield is strongly correlated with the density of spot defects. Therefore, estimation of functional yield has been reported to be adequate enough to predict manufacturing yield for such products as static and dynamic RAM memories. On the other hand, experience with bipolar IC's may lead to the conclusion that spot defects play an insignificant role as far as yield is concerned and that estimation of parametric yield is adequate to predict manufacturing yield. We wish to emphasize that while such simplified approaches to manufacturing yield prediction are possible in certain instances, they are generally unrealistic for VLSI technology.

To see this we consider three specific examples of VLSI chips: a moderate size CMOS gate array, a high-precision bipolar operational amplifier, and a large (e.g. 256 kbit) NMOS dynamic RAM. We consider the primary causes of yield loss for these IC's and illustrate the importance of various yield estimates. To compare the dominating failure modes and the main causes for yield losses in these IC's, we present the following figures. According to our previous classification, we have divided failure modes into two groups: structural and performance. The causes for failure have been divided into geometrical and electrical, and each of these is divided into global and local. To provide additional insight into the yield loss mechanisms in these IC's, we also specify which yield components (for both manufacturing and design yields) are affected the most.

A CMOS gate array is an example of an IC where high manufacturing yield is quite common. Due to relaxed layout design rules and very conservative design of functional blocks, soft performance failures are perhaps the most common. One of the main reasons for yield loss is an excessive signal delay due to long interconnections. Hence, estimation and prediction of parametric yield seems to be the only yield related problems of importance in the design and manufacture of gate arrays. Fig. 1-11 illustrates the relation between faults and disturbances for this type of circuit.

For the case of bipolar op amps, which are relatively small size IC's, yield is primarily sensitive to vertical effects due to relaxed layout design rules. More specifically, in bipolar IC's vertical impurity distributions almost

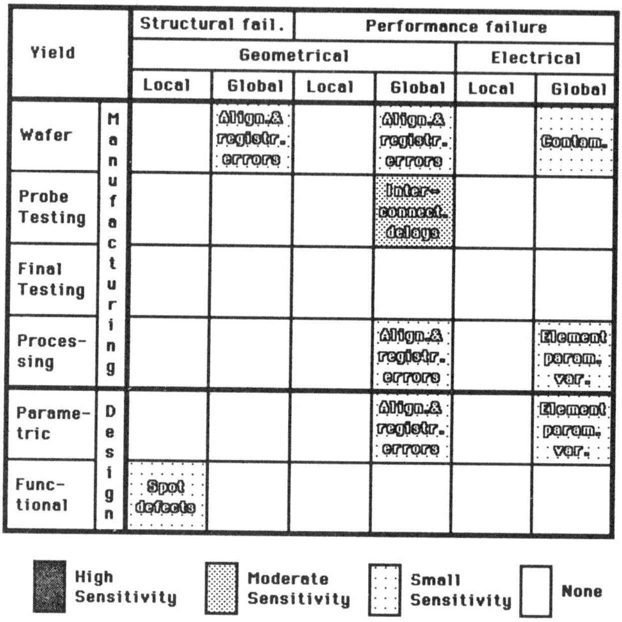

Figure 1-11: Yield sensitivity of a CMOS gate array.

exclusively determine the important device parameters such as current gain coefficients and resistances, and hence op amp performance. Thus, as summarized in Fig. 1-12, the wafer yield loss is primarily due to global variations in impurity profiles, such as improper junction depths and excessive emitter dip-effect. These faults are typically detected by junction depth measurements and may lead to rejection of wafers or even entire lots. Variations in impurity profiles and surface quality is also one of the most important reasons for a low probe testing yield. Other reasons include such local defects such as diffusion spikes or pipes which may led to shorts between emitter and collector regions, or creation of regions with significantly reduced base width that are susceptible to second breakdown effects. Soft performance faults can be also caused by mismatch in device parameters caused by linewidth variations due to lithographic processes. Observe that, in this type of IC, final tests are almost identical to probe tests, so that the percentage of chips rejected after final assembly should be extremely small. Also note that, for this case, test structure measurements can be used efficiently to diagnose the op amp failure modes, and therefore, the processing yield is quite similar to probe testing yield. Finally, as suggested in Fig. 1-12, parametric yield losses due to device parameter variations are more significant than functional yield losses. Therefore, it is critical to "center" the op amp design (see a discussion of design centering in Chapter 2) to make it less sensitive to the inherent process fluctuations. Hence, for the design and manufacturing of advanced bipolar circuits, both parametric and functional yield estimates may be required.

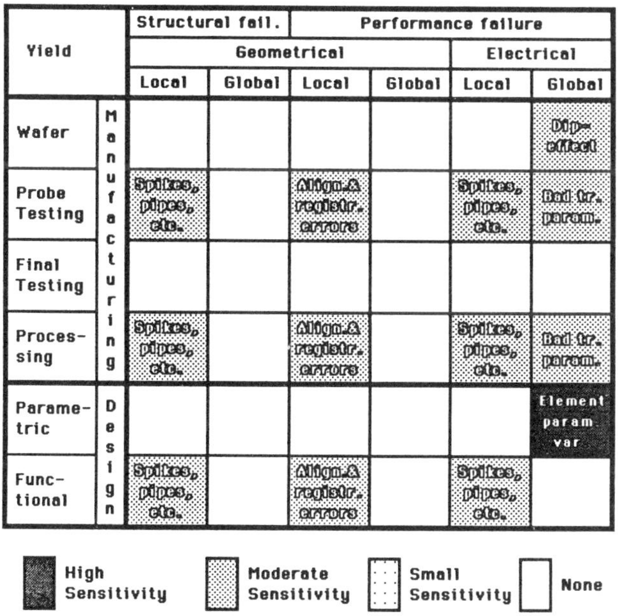

Figure 1-12: Yield sensitivity of an operational amplifier.

The relation between yield losses and the causes of such losses, for a large dynamic RAM is illustrated in Fig. 1-13. Of course, the manufacturing yield of such an IC is much lower than yield which can be achieved for the IC's discussed above. The dominating failure mechanisms for such VLSI circuits are also different. In order to achieve the density required to implement large dynamic RAM's, minimum device sizes and very aggressive spacing rules have to be employed. This results in large yield sensitivity to spot defects. Thus, structural failures that are caused by spot defects are the dominant factors which degrade yield. Usually these faults are detected by probe tests, and are either repaired, if there is sufficient redundancy on the chip, or the chips are rejected to avoid costly assembly operations. Thus

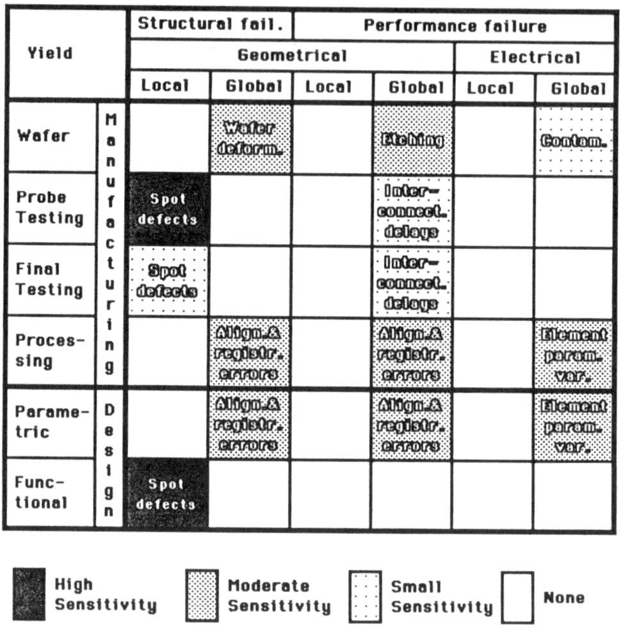

Figure 1-13: Yield sensitivity of a large dynamic RAM.

probe testing yield losses have been marked in Fig. 1-13 as the most significant. It has to be noted, however, that an IC of such complexity is sensitive to practically all process steps and global variations introduced in the process. Therefore during memory fabrication, careful in-line testing is performed to detect excessive variations in etching, wafer deformations or ion contamination. Hence, some wafers are rejected during the process and wafer yield is affected. Another interesting failure mode is soft parametric failure due to insufficient speed in either the write or read cycle. This failure can be caused by either excessive interconnect parasitics or incorrect device parameters (such as transconductance). In very fast dynamic memories, this failure mode seems to be very important. Hence we see that for the case of advanced memory circuits most types of process disturbances can contribute to yield loss and for effective design, as well as manufacturing, all types of failure modes and yields have to be considered.

Again, the above discussion has been presented to illustrate the point that while it has been the case that in less advanced technologies, or where there are relaxed design rules, yield losses are dominated by a single mechanism and a simplified approach to yield analysis can be employed. However, for VLSI circuits yield is determined by multiple phenomena. Therefore, an approach to yield analysis that combines various yield estimation methods is unavoidable.

1.5.3. Yield and Production Cost

The yield measures introduced in this chapter have been typically used for yield maximization purposes. The oldest and most practically sound approaches were experimentally-based and have been applied for maximization of both manufacturing and probe yields. Yield improvements have been achieved by changing process controls, IC layout or minimizing variability of the process disturbances. Although a complete discussion of these approaches is beyond the scope of this chapter, it is important to note their limitations. Specifically, these approaches attempt to improve process efficiency using yield as the sole criterion and thus overlook overall manufacturing process strategy. In fact, process efficiency has to be viewed from a wider perspective than just yield maximization. To see this consider the fabrication cost minimization problem.

The objective of an IC fabrication process is to produce a given number of working IC's within some given time period. To maximize profit, these IC's must be fabricated at minimal cost. Naturally, the higher the manufacturing yield, the lower the production cost per chip. However, since yield does not take into account the time and cost associated with manufacturing operations, nor the constraints on the number of chips that have to be manufactured within certain time period, it is not sufficient to predict costs. In fact, any cost function should include such factors as a penalty for not meeting the demand for a specified number of chips and the cost of operation and maintenance of the fabrication equipment, in addition to yield. These considerations are even more complicated if the cost of rework of selected manufacturing steps is taken into account. Such a complex cost function is necessary to perform a rigid statistical quality control and cost minimization process. However, since our goal here is to demonstrate the relationship between the production cost per chip and manufacturing yield, we will restrict our attention to much simpler expressions for the cost function.

To derive an expression for the overall cost of manufacturing W wafers (each of which contains N chips), we have to take into account the cost of fabricating the chips that went through the entire fabrication process (up to the final selection tests) as well as the cost of fabricating those chips (or wafers) that were rejected at intermediate inspection points. We note that the cost, C_1, of producing the $(W-W_W)$ wafers that are rejected during wafer processing is

$$C_1 = (W-W_W)(C_P^* + C_W) \tag{1.34}$$

where C_W is the cost of a wafer and C_P^* is the equivalent processing cost per wafer taking into account that wafers may be rejected at different stages of the wafer processing phase. Since it is more convenient to carry out an analysis in terms of chips rather than wafers, C_1 can be expressed equivalently as:

$$C_1 = (N - N_W)(c_P^* + c_W) \tag{1.35}$$

where the costs are specified on a per chip basis.

Similarly, the cost of producing the $(N_W - N_P)$ chips that are rejected after probe test, C_2, is given by:

$$C_2 = (N_W - N_P)(c_{PT} + c_P + c_W) \tag{1.36}$$

where c_{PT} is the probe testing cost per chip and c_P is the actual cost of wafer processing operations (including material cost, and equipment operation cost and maintenance) per chip.

Finally, the cost of producing the N_P chips that reach final test, C_3, is given by:

$$C_3 = N_P(c_{FT} + c_A + c_{PT} + c_P + c_W) \tag{1.37}$$

where: c_{FT} is the cost of the final test per chip and c_A is the overall assembly cost per chip.

Hence, the entire cost of manufacturing N chips is equal to:

$$C = N_P(c_{FT} + c_A + c_{PT} + c_P + c_W) +$$
$$N_W(1 - \frac{N_P}{N_W})(c_{PT} + c_P + c_W) + N(1 - \frac{N_W}{N})(c_P^* + c_W) \tag{1.38}$$

Note that only N_F out of N_P chips are fault-free and can be sold to customers. Therefore, the overall cost, C, divided by the number of good chips, N_F, is a good measure of fabrication process efficiency and should be the quantity which is to be minimized:

$$\frac{C}{N_F} = \frac{N_P}{N_F}(c_{FT} + c_A + c_{PT} + c_P + c_W) + \tag{1.39}$$
$$\frac{N_W}{N_F}(1 - \frac{N_P}{N_W})(c_{PT} + c_P + c_W) + \frac{N}{N_F}(1 - \frac{N_W}{N})(c_P^* + c_W)$$
$$= (c_{FT} + c_A + c_{PT} + c_P + c_W) + \frac{N_P - N_F}{N_F}(c_{FT} + c_A +$$
$$c_{PT} + c_P + c_W) + \frac{N_W}{N_F}(1 - Y_{PT}^*)(c_{PT} + c_P + c_W) +$$
$$\frac{N}{N_F}(1 - Y_W^*)(c_P^* + c_W)$$

$$= (c_{FT} + c_A + c_{PT} + c_P + c_W) + \frac{Y_P^*}{Y_M^*}(1 - Y_{FT}^*)(c_{FT} + c_A +$$
$$c_{PT} + c_P + c_W) + \frac{Y_W^*}{Y_M^*}(1 - Y_{PT}^*)(c_{PT} - c_P + c_W) +$$
$$\frac{1}{Y_M^*}(1 - Y_W^*)(c_P^* + c_W)$$

The above equation indicates that while the cost per good chip decreases if the manufacturing yield increases, the problems of cost minimization and yield maximization are not equivalent. For example, if the wafer yield loss is critical, better process management and supervision are necessary. If the probe yield loss is a concern, more stringent wafer rejection criteria based upon in-line measurements should be imposed (i.e., more adequate quality control system is necessary). Finally, if the yield loss during final test is significant, one of two conditions is present: either the fault coverage of the probe tests is not sufficient or the assembly operations have not been performed correctly. In the first case, the probe testing program has to be expanded, in the latter case, assembly operations have to be modified.

The transition from this cost function to profit maximization is trivial, since the profit per chip can be expressed as the difference between the market value of the IC chip and the cost of fabricating the chip. Hence, even such simple considerations can be useful for market analysis, production planning and scheduling. However, as we stated in the beginning of this section, additional factors such as the time required to satisfy demand for a particular chip and the penalty for not meeting the demand should enter profit considerations.

The above discussion indicates that maximization of individual yield components does not necessarily lead to an improvement in overall process efficiency. It is also important to note that to realistically solve the production cost minimization problem, a general framework for viewing process efficiency, such as presented in this chapter, is necessary.

1.6. Overview of the Sequel

In this chapter we have presented a framework that is useful for viewing the manufacturing yield of VLSI circuits. We have reviewed and classified the disturbances that cause yield loss as well as the effects that these disturbances have on performance. A number of yield measures that are useful both during the design and manufacturing processes have been defined and we have established relationships between these yield measures. Finally, we related yield to manufacturing costs for the purpose of providing a common denominator for the discussion of manufacturing process efficiency.

We now draw some general conclusions. The most important of these are:

1. The random phenomena that occur in an IC manufacturing process must be taken into account during the design and testing phases of IC realization. Towards this end, the basic physical nature of such random phenomena have to be investigated and

modeled in order to provide process and circuit designers with adequate information about process instabilities.

2. Both functional and parametric yields are important when determining process efficiency. While it has been the case that for some classes of circuits, one of these has dominated, in VLSI circuits we can not, in general, make such simplifications without exploring the principle causes of yield losses.

3. Both the IC manufacturing process and the IC design should be optimized with respect to a fabrication efficiency that can be expressed in terms of appropriately defined yields.

In the next chapter we introduce a technique that can be employed for maximizing parametric yield. This method is based upon approximating the region of acceptability with a polytope. By using such an approximation, the yield maximization problem becomes a linear program.

Evaluation of parametric yield during the design process relies on the ability to predict the effects of process disturbances on device parameters. In Chapter 3 we describe the simulator FABRICS which takes as input a set of process controls and device layout, and generates device model parameters compatible with the device models implemented in SPICE. To obtain accurate simulation, FABRICS II must be "tuned" to a particular IC fabrication process which requires the determination of a set of process disturbances. The program PROMETHEUS that accomplishes this task is also described. Finally, a complete system that can be used for process synthesis, called the Process Engineer's Workbench, is described.

Given the ability to simulate the fluctuations in the manufacturing process, it is possible to address a variety of issues. Hence, in Chapter 4 we introduce efficient methods for statistical timing simulation and worst case analysis. We also describe a method for optimizing the yield for minicells.

Chapter 5 considers the prediction of functional yield. A simulation tool that employs Monte Carlo method to determine the effect of spot defects on IC layout is discussed and used for yield estimation and redundancy analysis. We also describe a technique that evaluates the probability of failure caused by global deformations of the layout combined with spot defects effects. With the above models, it is possible to develop design rules that can maximize functional yield.

We conclude in Chapter 6, with a discussion of how the various tools and techniques introduced into the first five chapters can be used to develop a computer-aided manufacturing system.

Chapter 2
Parametric Yield Maximization

2.1. Introduction[1]

As indicated in the previous chapter, parametric yield is a function of the nominal values of the process controls and or the layout. Thus we can consider choosing a nominal design point that maximizes the number of circuits which meet a given performance specification. This approach to yield maximization is called *design centering* . Prior to 1977, Monte Carlo methods [48], [28], albeit crude and expensive, had been the technique primarily used for design centering. Bandler *et al.* [6] then proposed a design centering procedure that was directly related to classical nonlinear programming methods [33]. This approach was based upon selecting for each individual problem, a scalar objective functional which when minimized, inscribed a hypercube in the "feasible region" R of the n-dimensional parameter space.[2] However, the nonlinear programming approach does not attempt to approximate the boundary of the feasible region, but only determines points on this boundary. As will be seen below, an explicit approximation of the boundary of the feasible region is useful for solving a variety of other statistical design problems [89, 121, 16] as well as allowing one to obtain a Monte Carlo estimate of the yield at a reduced computational cost.

---

[1]This section is based upon the paper "The Simplicial Approximation Approach to Design Centering" by S. W. Director and G. D. Hachtel, *IEEE Transactions on Circuits and Systems*, Vol. CAS-24, No. 7, pp. 363-372, July 1977.

[2]The feasible region, or region of acceptability, R, is an n-dimensional subspace of the parameter space P which contains all points $p \in P$ which result in circuits that meet the desired specifications.

The simplicial approximation approach is a simple, efficient method for design centering that constructs an approximation to the feasible region and, therefore, is particularly well suited for maximizing the yield of integrated circuit designs. Specifically, the simplicial approximation method is based on explicitly approximating the boundary, ∂R, of the feasible region, R, of an n-parameter design space by a polyhedron made up of n-dimensional simplices[3]. Approximation is necessary since ∂R is generally known only in terms of nonlinear inequality constraints that express acceptable circuit performance in terms of voltages and currents which depend implicitly upon the design parameters.[4] It is assumed that the constraint functions are locally convex, i.e., that the sequence of points generated on ∂R are extreme points of a convex set. The case where this assumption is violated involves mathematical representation of a nonconvex polyhedron, but can still be handled, as indicated below.

The simplicial approximation approach may also be generalized to directly solve a variety of statistical design problems such as worse-case tolerance assignment problems with arbitrary (e.g. asymmetrical and/or mixed independent and joint) distributions, and mixed worst-case and yield maximization problems. The latter problem arises when some of the design parameters are statistical in nature while other parameters, such as power supply voltages, are worst case. In other words, the circuit is required to operate satisfactorily for all values of power supply voltages within some interval. Such generalization will be considered after the basic technique has been introduced.

2.1.1. Definitions of Yield and Design Center

In the remainder of this chapter we assume that the network under consideration obeys a set of differential-algebraic equations of the form

$$g(\xi,\dot{\xi},x,t)=0, \qquad 0 \le t \le t_f \tag{2.1}$$

---

[3] Butler, in [16], was one of the first to try and approximate the feasible region, however he was interested in graphical displays of the region and therefore approximated only projections of this region onto two dimensional subspaces.

[4] It is important to recognize that determination of the voltages or currents in a circuit for a given set of design parameters requires a complete circuit simulation, i.e., the solution of a set of nonlinear differential-algebraic equations.

where ξ is a vector of node voltages, branch voltages, and branch currents and x is an n-vector of statistically varying parameters (for example process controls, layout parameters, or process disturbances) with a joint probability density function (jpdf).

$$f_x(x,\mu_x,\Sigma_x) \tag{2.2}$$

More will be said about the choice of design parameters in Sec. 2.5 where μ_x is the mean value of x and Σ_x is the covariance matrix.[5] Further, let X denote the n-dimensional parameter space so that $x \in X$. We will consider a circuit with parameter values x to be acceptable if the solution $\xi(t)$ of Eq. (2.1) obeys a given vector of n_c constraints, whose components are defined by

$$\Phi_i(x)=\int_0^{t_f}\phi_i(\xi,\dot{\xi},x,t)\;dt \leq 0, \qquad i=1,2,\cdots,n_c \tag{2.3}$$

In addition the parameter vector x is assumed to lie between some lower and upper limits, or *box constraints*

$$x^l \leq x \leq x^u \tag{2.4}$$

Note that the constraints Φ_i depend on x implicitly through the solution $\xi(t)$ of Eq. (2.1) as well as through its role as an explicit argument of the Φ_i. The set of constraints defines the *region of acceptability* R. In Chapter 1 we denoted this region by A_X. However, for the presentation in this chapter, this notation is more convenient.

$$R=\{x|\Phi_i(x) \leq 0, \text{ for all } i \in [1,2,\cdots,n_c] \text{ and } x^l \leq x \leq x^u\} \tag{2.5}$$

Note that R is closed, and is bounded if the limits in Eq. (2.4) are finite. Of particular interest will be the boundary of R, denoted by ∂R and defined by

---

[5]Until stated otherwise, we assume that all design variables are normally or log normal random variables and all disturbances are characterized in terms of deviations of the designable parameters from their nominal value.

$$\partial R=\{x|\Phi(x) \leq 0,\ x^l \leq x \leq x^u, \text{ and } \Phi_i=0, \\ \text{or } x_i=x_i^l, \text{ or } x_i=x_i^u, \text{ for some } i\} \tag{2.6}$$

The yield associated with some nominal design point, $\mu_x = X_0$ is

$$Y= \text{prob}\,(x \in R)=\int\int_R \cdots \int f_x(x,X_o,\Sigma_x)\,dx \tag{2.7}$$

where the integration in Eq. (2.7) is over the entire feasible region R.

The purpose of design centering can now be stated as that of choosing the nominal values X_0 of the designable parameters so that yield Y is maximum for a given distribution $f_x(X, X_o, \Sigma_x)$. Since, in general, the distributions are known and fixed, the design centering problem can be defined to have two clearly delineated phases, namely

1. determination of the feasible region R and the design center, and

2. evaluation of the yield integral Eq. (2.7).

2.1.2. Design Centering By Simplicial Approximation

The method of simplicial approximation is based on approximating the boundary ∂R of the feasible region R by a polyhedron, i.e., by the union of those portions of a set of n-dimensional hyperplanes which lie inside ∂R or on it. The procedure begins by determining any $m \geq n+1$ points, $x_1, x_2 \cdots x_m$ on the boundary ∂R. Usually m is taken to be either $n + 1$ or $2n$. One way to find $2n$ points is by first finding a feasible set of designable parameters inside R and then performing one-dimensional line searches in the positive and negative coordinate directions. The initial feasible point can be obtained either by designer insight or through the use of an optimization program. Given a set of m points on R, the convex hull (polyhedron) of these points is then constructed.[6] The convex hull is characterized by the set of m_H inequalities

$$\eta_k^T x \leq b_k, \quad k=1,2, \cdots, m_H \tag{2.8}$$

where η is a unit "outward pointing," vector normal to the kth hyperplane defined by

$$\eta_k^T x = b_k, \quad k=1,2, \cdots, m_H \tag{2.9}$$

and b_k is a measure of the distance of the kth hyperplane from the origin. Moreover, the convex hull of the set of points x_j, $j=1,2, \cdots, m_H$ approximates ∂R in an *interior* sense. That is, under our assumption of convexity of R, every point interior to these boundary of hyperplanes is also inside ∂R.

Given the first approximation to ∂R we can find a first estimate of the design center by determining the center of the largest hypersphere which can be inscribed inside the polyhedron of defined by the m_H hyperplanes described by Eq. (2.9). The center of the largest hypersphere which can be inscribed within this polyhedron can be found quite easily using a linear programming approach. First recognize that the distance from a point x^* inside the polytope to the kth hyperplane is

$$d_k = \eta_k^T x^* - b_k \tag{2.10}$$

Therefore the center and radius of the largest hypersphere can be found by determining the maximum value of r and the point x^* for which

$$\eta_k^T x^* + r \leq b_k, \quad k=1,2, \cdots, m_H \tag{2.11}$$

---

[6]The convex hull can easily be constructed by considering the hyperplanes defined by combinations of the m boundary points taken n at a time. All hyperplanes which have all boundary points to one side are part of the convex hull. There are more efficient procedures to determine the convex hull, however since this step is performed only once in the entire procedure, efficiency is not of prime importance here.

This linear program is the dual of the following standard (primal) linear programming problem. Minimize the objective function

$$\Phi = \sum_{k=1}^{m_H} b_k \lambda_k \tag{2.12}$$

subject to the constraints

$$\begin{bmatrix} \eta_1 & \eta_2 & \cdots & \eta_{m_H} \\ \downarrow & \downarrow & & \downarrow \\ & \cdot & & \cdot \\ 1 & 1 & \cdots & 1 \end{bmatrix} \begin{bmatrix} \lambda_1 \\ \lambda_2 \\ \cdot \\ \cdot \\ \lambda_{m_H} \end{bmatrix} = \begin{bmatrix} 0 \\ 0 \\ \cdot \\ \cdot \\ 1 \end{bmatrix} \tag{2.13}$$

and $\lambda_k \geq 0$ for all k. The primal form of this linear programming problem is more desirable because the number of unknowns is less than the number of constraints. Any standard linear programming computer subroutine can be used to solve Eq. (2.12) and Eq. (2.13) and as a byproduct yield the results of the dual problem, i.e., the radius r and center x^* of the largest inscribed hypersphere.

It is important to recognize the fact that the computational effort required to determine the convex hull of a set of boundary points and then inscribe a hypersphere in this convex hull is usually considerably less than the computational effort required to perform a nonlinear transient circuit simulation. In fact, it is reasonable to assume that the computational cost of performing a circuit simulation far exceeds the computational cost of solving any linear program or of performing any of the matrix-vector manipulations required by this method.

The next step in the design centering procedure is to improve the simplicial approximation to ∂R and then update the design center. This step is accomplished by determining which of the m_H "faces" of the polyhedron (i.e., which of the m_H truncated hyperplanes defined by Eq. (2.9)) that are tangent to the largest inscribed hypersphere is the "largest." The largest tangential face is of particular interest since it is the one which is most likely to be the poorest approximation to ∂R. The faces of the polyhedron which are tangent to the largest inscribed hypersphere are those associated with the "zero slack variables" used in solving the linear program (see Eq. (2.12) and Eq. (2.13)). The largest tangential face is defined as the face in which the largest $(n\text{ -}1)$-dimensional hypersphere may be inscribed.

The procedure for finding the largest hypersphere which can be inscribed in a face of the approximating polyhedron is only a slight variation of the linear programming procedure outlined above for finding the largest hypersphere inscribed in the polyhedron itself. Suppose we wish to find the largest hypersphere inscribed in the jth face of the polyhedron. The center of this hypersphere, denoted by x_j^*, must be on the jth hyperplane so that

$$\eta_j^T x_j^* = b_j \tag{2.14}$$

Now let $\eta_j^\perp$ denote a unit vector perpendicular to η_j, i.e.,

$$(\eta_j^\perp)^T \eta_j = 0 \tag{2.15}$$

Thus $\eta_j^\perp$ lies in the jth hyperplane and the surface of a hypersphere of radius r_j centered at x_j which lies in the jth hyperplane is described by

$$x_j^* + r_j \eta_j^\perp \tag{2.16}$$

The largest such hypersphere is the one with as large an r_j as possible subject to the constraint that all points on the surface of this hypersphere lie within the approximating polyhedron, i.e., they satisfy the constraints

$$\eta_k^T (x_j^* + r_j \eta_j^\perp) \le b_k, \qquad \text{for } k=1,2,\ldots,m_H; \quad k \ne j \tag{2.17}$$

which can be shown to be equivalent to the constraints

$$\eta_k^T x_j^* + r_j \sin\phi_{jk} \le b_k \tag{2.18}$$

where ϕ_{jk} is the angle between η_k and η_j. Thus the linear program to be solved is to determine x^* and maximize r subject to Eq. (2.15) and Eq. (2.18). Note that Eq. (2.15) can be expressed as

$$\eta_j^T x^* \leq b_j \tag{2.19}$$

and

$$-\eta_j^T x^* \leq -b_j \tag{2.20}$$

so that this linear program is the dual form of the following standard (primal) linear programming problem.

Minimize the objective function

$$\Phi = \sum_{k=1}^{m_H+1} b_k \lambda_k \tag{2.21}$$

subject to the constraints

$$\begin{bmatrix} \eta_1 & \eta_2 & \cdots & \eta_j & \cdots & \eta_{m_H} & -\eta_j \\ \downarrow & \downarrow & & \downarrow & & \downarrow & \downarrow \\ \sin\theta_{1j} & \sin\theta_{2j} & \cdots & 0 & \cdots & \sin\theta_{m_H j} & 0 \end{bmatrix} \times \begin{bmatrix} \lambda_1 \\ \lambda_2 \\ \cdot \\ \cdot \\ \cdot \\ \lambda_{m_H+1} \end{bmatrix} = \begin{bmatrix} 0 \\ 0 \\ \cdot \\ \cdot \\ \cdot \\ 1 \end{bmatrix} \tag{2.22}$$

and $\lambda_k \geq 0$ for all k. Note that in Eq. (2.21) $b_{m_H+1} = -b_j$.

The above procedure is repeated for each tangential face, the largest face being the one with the largest inscribed hypersphere.[7] Once the largest face is determined, a one-dimensional search is made in the outward normal direction, starting from the center of the largest inscribed hypersphere in that face, for a new point on the boundary. This new point is then added to the set of boundary points previously found and a new convex hull is found.[8] Thus our approximation to ∂R has been improved and a new design center can be found. The entire procedure can then be repeated.

2.1.3. Design Centering Procedure

The proposed design centering procedure , illustrated for two dimensions in Fig. 2-1, is as follows.

Step 1 Determine a set of $m \geq n + 1$ points on the boundary, ∂R.

Step 2 Find the convex hull of these points and use this polyhedron as the initial approximation to ∂R. Set $k = 0$.

Step 3 Inscribe (using the linear program Eq. (2.22)) the largest hypersphere in this approximating polyhedron and take as the first estimate of the design center the center of this hypersphere.

Step 4 Find (using the linear program Eq. (2.12)), the "mid-point" of the "largest" face in the polyhedron which is tangent to the inscribed hypersphere.

---

[7] In general, for an n-dimensional parameter space, there will be $n + 1$ tangential faces. Thus there are about $n + 1$ linear programs that must be solved, which is much fewer than if the size of all the faces of the polyhedron had to be found.

[8] Determining the new convex hull at this point requires two steps. First, all faces of the original polytope for which the defining inequality Eq. (2.8) no longer holds when the new boundary point is considered, must be deleted. Next hyperplanes formed by combinations of $(n - 1)$ points taken from the set of points which define the deleted faces and the new boundary point are tested to see if all boundary points are on one side. If this test is passed, the hyperplane is part of the convex hull, if not it is ignored.

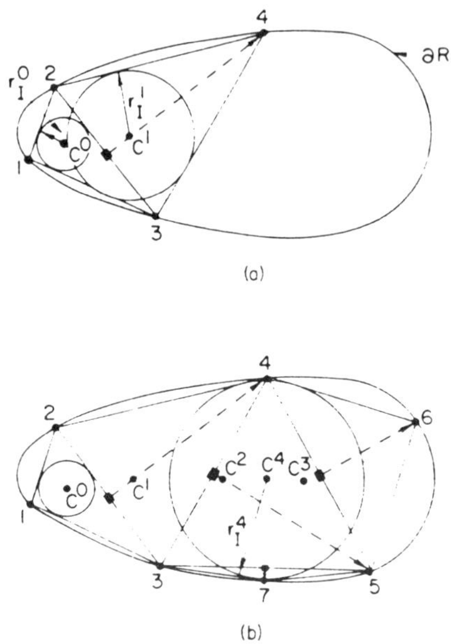

Figure 2-1: Illustration of simplicial approximation.
(a) Initial polyhedron defined by points (1,2,3) and breaking the largest face (2-3).
(b) The resulting polyhedron after four inflation iterations.

Step 5 Find a new boundary point[9] on ∂R by searching along the outward normal of the "largest" face found in step (4) extending from the "midpoint" of this face.

Step 6 Inflate the polyhedron by forming the convex hull of all previous points plus the new point generated in step (5).

Step 7 Find (using the linear program Eq. (2.12)), the center p_c^k and radius r_I^k of the largest hypersphere inscribed in the new polyhedron found in step (6). Set $k = k + 1$. Go to step (5).

[9] It can occur that the largest face actually lies in ∂R. We test for this case, and if it occurs, the face is flagged, and never again considered as a candidate for step (4). In this case, step (4) is repeated using the next largest face.

Step (5) in the above procedure is key since it involves a line search along a given direction. Each step in this search involves first solving the circuit equations Eq. (2.1) and then evaluating the constraints Eq. (2.3) and Eq. (2.4) to see which, if any, are violated. The search ends when point x_B is located for which

$$\Phi_j(x_B)=0, \quad \Phi_i(x_B) \leq 0, \quad i \neq j. \tag{2.23}$$

The above procedure will rapidly generate a sequence of points whose convex hull forms an interior polyhedron, ∂R_I which approximates ∂R. Step (7), the process of inscribing the largest hypersphere in ∂R_I can be used as a means of monitoring the convergence of $\partial R_I \rightarrow \partial R$. If R is convex, as per our assumption, the sequence of centers $\{x_c^k\}$ will converge to the center of the largest hypersphere that can be inscribed in ∂R, and the associated sequence of radii, $\{r_I^k\}$, will converge to the radius of the largest inscribed hypersphere. This sequence is considered to have converged when

$$|r_I^{k+1} - r_I^k| \leq \epsilon_R r^k + \epsilon_A \tag{2.24}$$

where ϵ_R and ϵ_A are given relative and absolute convergence parameters. Convergence is expected on the basis of the following.

Theorem 1

1. Let r_I^k be the radius of the largest hypersphere inscribed in a polyhedron ∂R_I^k formed from $n+1+k$ points on a convex surface ∂R in n-space.

2. Let r^{k+1} be the radius of the largest hypersphere inscribed in a polyhedron ∂R_I^{k+1} formed by applying the inflation step of the design centering procedure to ∂R_I^k, then

$$r_I^{k+1} \geq r_I^k \tag{2.25}$$

3. Further, if the outward normals to the faces of ∂R_I^k are not parallel and if the inscribed hypersphere is tangent to only $n + 1$ faces, then

$$r_I^{k+1} > r_I^k \tag{2.26}$$

Note that condition 3 of Theorem 1 is not restrictive in practice because if it is violated, a finite number of passes through the design centering procedure will cause condition 3 to be satisfied. For example, if two tangential faces are parallel, breaking the largest face may not allow the sphere to grow. However, upon reapplication of the procedure often enough (no more than n reapplications are necessary), then one of the parallel faces will eventually be broken. Thus we have the following.

Corollary

$$r_I^{k+1+l} > r_I^k, \quad 0 < 1 \leq n \tag{2.27}$$

Assume that there are no unsuspected surface pathologies which would cause the simplicial approximation process to converge to some limit *inside* ∂R.[10] Then it would follow from the above theorem and corollary that the approximating simplex ∂R_I^k would, as $k \rightarrow \infty$, fill ∂R like a ball being inflated by an air pump. Theorem 1 can be proven by observing that the new polyhedron contains the old polyhedron. The corollary follows from the fact that if two faces are parallel, breaking one of them destroys this relationship.

A word about the computational effort required to use the above procedure is in order. Our basic premise is that this procedure is to be used in situations where the number of statistically varying design parameters is small compared to the number of the algebraic-differential equations that need to be solved in order to determine circuit behavior. This situation is common for the design of integrated circuits where the variations in circuit

[10] Theorem 32 of [27] shows that the above procedure will produce a convergent subsequence as $k \rightarrow \infty$. Theorem 33 of [27] shows that ∂R can be approximated by a polyhedron to arbitrary accuracy. However, we have as yet no proof that $\partial R^k \rightarrow \partial R$ as $k \rightarrow \infty$. This has not been a problem in practice.

parameters are related to the variations in a much smaller number of processing parameters. For these cases, the dimensionality of the linear programs that are solved in the above procedure is relatively small when compared to the dimensionality of the circuit simulation problem. Thus the time spent in generating and inflating the convex hull is small when compared to the time involved in performing a transient simulation of a large nonlinear network. Based upon the above reasoning the most expensive operation in the above procedure is the search in step (5) to find a new boundary point. Therefore the search step must be efficient.

Before leaving this section we note that with each interior approximation ∂R_I^k may be associated a yield

$$Y_I^k = \int \int_{R_I^k} \cdots \int f_x(x, \mu_x, \Sigma_x) dx \tag{2.28}$$

where R_I^k is the volume in n-space bounded by ∂R_I^k, i.e, the kth-interior approximation to R. Note also that since ∂R_I^k consists of interior bounding hyperplanes

$$Y_I^{k-1} < Y_I^k \leq Y \tag{2.29}$$

where Y is computed over the true feasible region R. That is, the interior simplicial approximation gives a lower bound on the yield. Moreover calculation of this yield requires only generation of random sets of parameter values and determination of the number which fall inside ∂R_I, i.e., only a Monte Carlo analysis on the input space is required: no multitude of circuit simulations is necessary. (For further discussion of these ideas see [23].)

2.1.4. Scaling: Inscribing a Hyperellipsoid

If the design centering procedure outlined in the previous section is employed when the region of acceptability is rectangular in shape, the best possible results may not be obtained. For elongated regions of acceptability, it would be more appropriate to determine the design center by inscribing a hyperellipsoid rather than a hypersphere. Fortunately this result is easily obtained by merely scaling the parameters prior to employing the design centering procedure. One scheme for scaling, which has proven to be successful, is based upon using an estimate of the spreads in possible parameter values. We now describe this technique.

Assume that the mth step of the design centering procedure has been carried out and that $n+m+1$ boundary points $\{x_k\}$, $k=1,2,\cdots,n+m+1$ have been found. The lower and upper bounds of each parameter are now determined

$$x_i^L = \min_k \{x_{ik}\} \tag{2.30}$$

and

$$x_i^U = \max_k \{x_{ik}\}$$

for $i=1,2,\cdots,n$ where x_{ik} is the ith coordinate of the kth-boundary point. The scale factor for the ith component of the parameter vector is given by

$$S_i = x_i^U - x_i^L, \quad i=1,2,\cdots,n \tag{2.31}$$

and a scale matrix is defined

$$\mathbf{S} = \text{diag}\,\{S_1, S_2, \cdots, S_n\} \tag{2.32}$$

A scaled set of boundary points are now defined

$$\hat{\mathrm{x}}_k = \mathbf{S}^{-1} x_k, \quad k=1,2,\cdots,n \tag{2.33}$$

and steps (3) and (4) of the design centering procedure can be carried out as before using the set of points $\{x_k\}$. Observe though that the radius of the largest hypersphere, $\hat{\mathrm{r}}$, found in step (3) can be unscaled to give the "radii," or axis lengths, of a hyperellipsoid

$$r_i = S_i \hat{\mathrm{r}}, \quad i=1,2,\cdots,n \tag{2.34}$$

which can be thought of as being inscribed in the unscaled polyhedron defined by the points $\{x_k\}$. Similarly, if $\hat{C}$ denotes the design center of the scaled problem found in step (3), the design center of the unscaled problem is

$$C = \mathbf{S}\hat{C} \tag{2.35}$$

This interpretation of C as the center and the r_i as the radii of a hyperellipsoid comes about by recognizing that in the scaled problem the largest hypersphere of radius $\hat{r}$ and center $\hat{C}$ is characterized by the locus of points $\hat{x}$ for which

$$\sum_{i=1}^{n} (\hat{p}_i - \hat{C}_i)^2 = \hat{r}^2 \tag{2.36}$$

or

$$\sum_{i=1}^{n} \frac{(p_i - C_i)^2}{S_i \hat{r}} = 1 \tag{2.37}$$

In terms of the unscaled space, Eq. (2.37) is the equation of a hyperellipsoid. Note that r_i is *not* the distance between the design center C and the ith face of the polyhedron (as it would have been without scaling) but the distance from the design center to the surface of the hyperellipsoid along the ith axis.

Clearly the results of the above procedure are dependent upon the scaling. It is suggested that once the design centering procedure has converged, the lower and upper bounds on the parameters be redetermined, considering now all the boundary points of ∂R_I^{m+1}, and the scaling adjusted. The design centering procedure should then be repeated to ensure that a good nominal point had been found.[11]

[11]Note that all boundary points previously found as well as the approximating polyhedron can be rescaled and used as the starting approximation here. Thus, in general, little additional work will be required to repeat the design centering procedure.

2.1.5. Illustration of the Basic Method

To illustrate the design centering algorithm as described above, consider the three transistor NAND gate circuit of Fig. 2-2. In order that we may graphically illustrate the steps in the procedure, we consider only three designable parameters: the width W_1 of the load device $T1$, the width W_{23} of the input devices T2 and T3 (both T2 and T3 have the same width), and the threshold voltage, V_T, of the three devices (the same for all three). The lengths of all three devices are held fixed. Therefore we have two geometric parameters and one processing related variable to adjust.

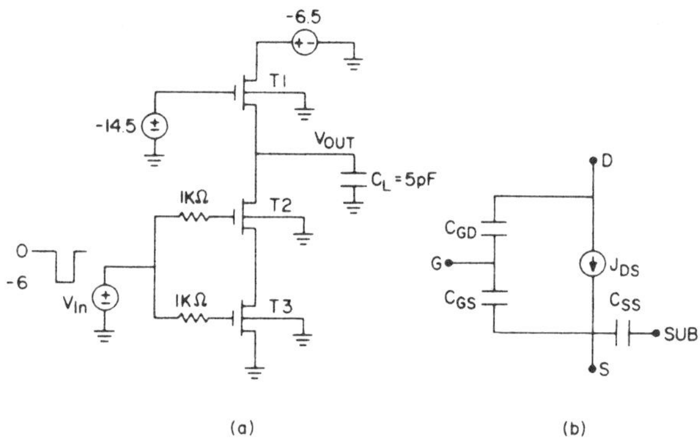

Figure 2-2: (a) Two input NAND gate used in the design centering example.
(b) The FET model used: $J_{DS} = G_m(V_{GS} - V_T)^2W/L$ above pinch-off and $J_{DS} = G_mV_{DS}(V_{GS} - V_T - V_{DS}/2)W/L$ below pinch-off, where W is width and L is length of device. Capacitances are functions of W and L: $C_{GS} = (L + \alpha)W\beta$; $C_{GO} = \alpha W\beta$; $C_{SS} = (W/\gamma + \delta)\epsilon$; where α, β, γ and ϵ are constants.

There are three constraints that must be satisfied in order for the circuit to be considered acceptable: the total area A, occupied by the three devices must be less than 2500 mils^2; the delay time, T_D (i.e., the length of time between which the input passes through $-3V$ and the output falls through $-3.25V$) must be less than 110 ns, and the output voltage V_0, be greater than $-0.7V$ when the gate is on. In addition we have the following *box* constraints on the parameters

$$5.0 \leq W_1 \leq 50.0 \quad \text{(microns)} \tag{2.38}$$

$$50.0 \leq W_{23} \leq 250.0 \quad \text{(microns)}$$

and

$$1.0 \leq V_T \leq 2.0 \quad \text{(volts)} \tag{2.39}$$

The feasible starting point is $W_1 = 12$, $W_{23} = 230$, and $V_T = -1.2$. For this set of parameter values, $A = 2490$, $T_p = 73.8$ and $V_0 = -0.558$.

The first step in the procedure is to find six points on the boundary of the feasible region by searching in the positive and negative coordinate directions. The six points found, and the associated constraint, values are shown in Table 2-1. The constraints that are satisfied exactly are framed. In two cases, the box constraints on V_T were the limiting constraints.

W_1	W_{23}	V_T	A	T_D	V_0
12.85	230	1.2	[2500]	68.5	-0.6
8.28	230	1.2	2442	[110]	-0.395
12	231.1	1.2	[2500]	73.8	-0.555
12	180.7	1.2	1989	70	[-0.7]
12	230	[2]	2490	78.7	-0.631
12	230	[1]	2390	72.8	-0.543

Table 2-1: Initial Boundary Points.

A sketch of the initial approximating simplex defined by these boundary points is shown in Fig. 2-3. The x axis corresponds to $W_1/4.6$, the y axis corresponds to V_T, and the z axis corresponds to $W_{23}/50.3$. The scaling was determined by the initial spread in parameter values as determined by the coordinate searches above. The axes emanate from the "center" of the body. (Note that this figure represents a perspective drawing and does not necessarily display the actual relative size of each face.) The first estimate of the design center, based upon the first approximating simplex, is $W_1 = 11.4$, $W_{23} = 219$, and $V_T = 1.32$. The radii (i.e., axis) of the largest inscribed ellipsoid (i.e., unscaled sphere radius) is 1.1 for W_1, 11.7 for W_{23}, and 0.23 for V_T. In other words, given this design center, acceptable circuits would be obtained for a deviation of 9.6 percent in W_1 if W_{23} and V_T were kept fixed, a 5.3-percent deviation in W_{23} if W_1 and V_T were kept fixed and a 17.4 percent deviation in V_T if W_1 and W_{23} were kept fixed. Note that since a hyperellipsoid is being inscribed in the region of acceptability, and not a hypercube whose sides are twice the maximum deviation, there is no guarantee that acceptable circuits will result if two or more parameters deviate maximally from the nominal.

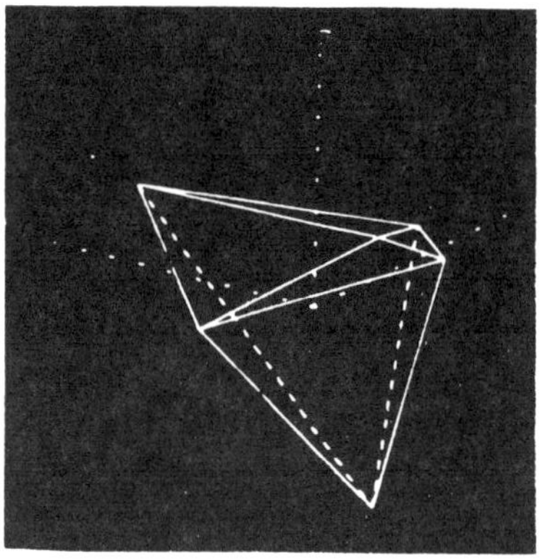

Figure 2-3: Initial (scaled) approximating polyhedron to region of acceptability for the NAND gate example. Axes emanate from an estimate of design center.

The next step involves searching for a new point on the boundary starting from the center of the largest face and proceeding in an outward normal direction. A new polyhedron is then constructed using this new point and a new estimate of the design center made. Fig. 2-4 shows the new polyhedral approximation. This procedure is repeated until no increase in the size of the inscribed hypersphere is possible. The final approximating polyhedron is shown in Fig. 2-5. The final design center is $W_1 = 10.6$, $W_{23} = 209$, and $V_T = 1.47$; and the radius of the inscribed hyperellipsoid is 2.2 for W_1, 24 for

W_{23} and 0.47 for V_T. Acceptable circuits would be obtained even if there was a 21 percent change in W_1 if W_{23} and V_T were kept fixed, a 11.5 percent change in W_{23} if W_1 and V_T were kept fixed, and a 32 percent change in V_T if W_1 and W_{23} were kept fixed. Hence, the design centering procedure increased the allowable tolerances of the designable parameters by a factor of two over the value quoted above for the first computed design center (which was itself an improvement over the initial feasible point).

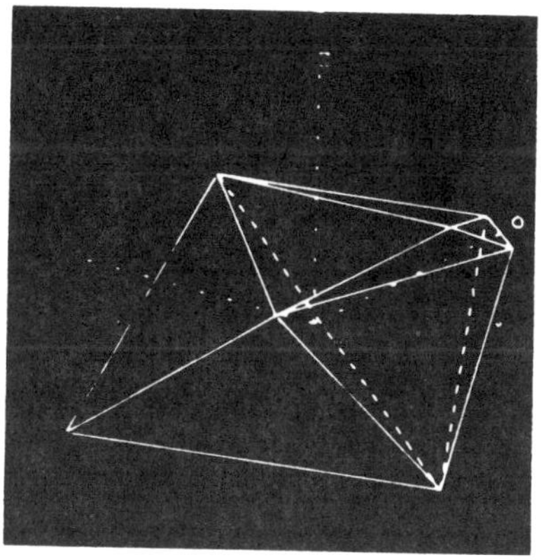

Figure 2-4: Resulting polyhedron when largest face of polyhedron in Fig. 2-3 was broken.

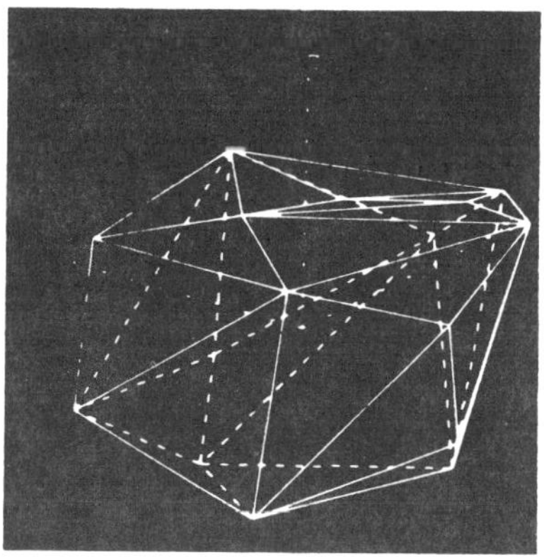

Figure 2-5: Final polyhedral approximation.

2.2. Design Centering and Worst Case Design with Arbitrary Statistical Distributions[12]

We now generalize the basic simplicial approximation method so as to be able to directly solve a variety of statistical design problems such as worst-case tolerance assignment problems with arbitrary distributions and mixed worst-case and yield maximization problems. In what follows it is assumed that all statistical parameter distributions have equal variance or spread. No loss of generality is incurred by this assumption, and we show how differences in variance may be handled by appropriate scaling of the norm. We also assume that the feasible region of the statistical design problem is convex. Thus any point which lies inside the simplicial approximation is a feasible point.

We begin our generalization of the simplicial approximation method in the next section by showing how norms can be associated with probability density functions in a natural way. The concept of a *norm body* is introduced enabling us in Section 2.2.2 to consider design centering for the case of arbitrary statistical distributions.

2.2.1. Norm Bodies and PDF Norms

In general, for arbitrary joint statistical distributions in n dimensions, a level contour on the surface of the pdf defines a closed body. In what follows, we consider only distributions where this closed body is well approximated by a convex body. To any such convex body we can associate a norm $n(x)$ with the properties[13]

1. $n(x) \geq 0$ for all x

2. $n(x+y) \leq n(x) + n(y)$

---

[12] This section is based upon the paper "Yield Maximization and Worst Case Design with Arbitrary Statistical Distributions", by R. K. Brayton, S. W. Director and G. D. Hachtel, *IEEE Transactions on Circuits and Systems*, Vol. CAS-28, No. 9, pp. 756-764, Sept. 1980.

[13] Generally a norm, $n(x)$, has the property that $n(\alpha x) = |\alpha| n(x)$, sometimes termed an absolute or equilibrated norm. In this paper the term *norm* denotes a larger class of functions in which $n(x) \neq n(-x)$ necessarily [114]. This means that asymmetrical distributions can be allowed in what follows.

3. $n(\alpha x) = \alpha n(x)$ for $\alpha > 0$.

In particular, the 2-norm

$$n_2(x) = \|x\|_2 = \left(\sum_{i=1}^{n} x_i^2\right)^{1/2} \tag{2.40}$$

whose level contours are n-spheres, can be used to characterize the level contours of an n-dimensional, equivariant, joint *normal* distribution with mean x_0, by defining the surface

$$n_2(x - x_o) = r \tag{2.41}$$

The associated closed convex body is defined by

$$\{x | n_2(x - x_o) \leq r\} \tag{2.42}$$

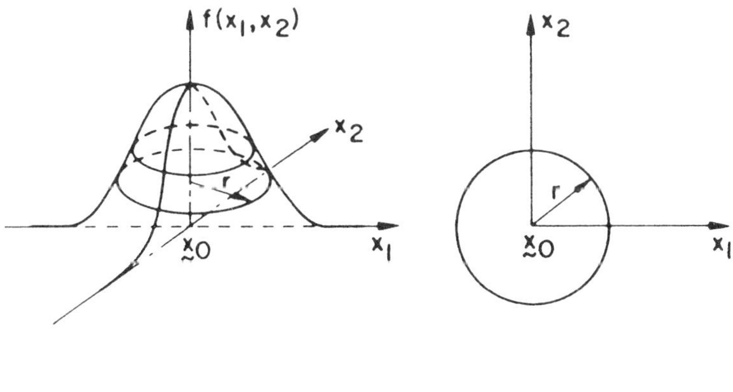

Figure 2-6: (a) Illustration of the level contour for a 2-dimensional equivariant, joint normal distribution.
(b) Projection of a level contour onto the parameter space.

and is illustrated for two dimensions in Fig. 2-6. Similarly, the max (or infinity) norm

$$n_{\infty}(x)=||x||_{\infty}=\max_{i}\{|x_i|\} \tag{2.43}$$

which corresponds to an n-cube, can be used to characterize the level contours of an n-dimensional, equal width, joint *uniform* distribution with a mean of x_0, by defining the surface

$$n_{\infty}(x-x_o)=r \tag{2.44}$$

The associated closed convex body is defined by

$$\{x|n_{\infty}(x-x_o)\leq r\} \tag{2.45}$$

and is illustrated for two dimensions in Fig. 2-6. A norm which is used to characterize the level contour of a jpdf is referred to as a *jpdf-norm* and the convex body associated with this norm will be referred to as a *norm body* [114].

In addition to the concept of norm bodies as defined above, we will need the concept of a dual norm. Associated with each norm *n(x)* there is a corresponding *dual norm* $n^*(x)$, [114], defined by

$$n^*(x)=\max_{y}\{y^Tx|n(y)\leq 1\} \tag{2.46}$$

While in general Eq. (2.46) is an optimization problem with a linear objective function and nonlinear convex constraint, it is well known that if

$$n_p(x)=||x||_p\equiv\left(\sum_{i=1}^{n}|x_i|^p\right)^{1/p} \tag{2.47}$$

the dual norm is

$$n_p^*(x)=||x||_q \tag{2.48}$$

where

$$\frac{1}{p}+\frac{1}{q}=1 \tag{2.49}$$

In other words, the dual norm of the 2-norm is the 2-norm, while the dual norm of the max- (or infinity) norm is the 1-norm

$$||x||_i = \sum_{i=1}^{n} |x_i| \tag{2.50}$$

and vice versa. Furthermore, for the case of an arbitrary norm body, we can approximate the norm body itself by a simplicial approximation, in which case Eq. (2.46) reduces to a linear program.

Of key importance in the development of the algorithm for the yield maximization procedure described in Section 2.1.3, and generalized below, is the ability to compute the distance from a point to a hyperplane in any arbitrary norm. The distance in norm $n(x)$ from a point x to a hyperplane π, is defined by [58]

$$d_n(x,\pi) \equiv \min_y \{n(y-x) | y \in \pi\} \tag{2.51}$$

The following theorem establishes that $d_n(x,\pi)$ is linear in x thus making it possible to easily extend the yield maximization procedure described earlier so as to handle arbitrary norm bodies.

Theorem 2 (Normed Distance to a Hyperplane)

Let $n(x)$ be a norm of x and $n^*(x)$ be the corresponding dual norm of x defined by Eq. (2.46). Then the distance in n-norm from point x_0 to hyperplane $\pi \equiv \{x | \eta^T x = b\}$ is

$$d_o \equiv d_n(x_o, \pi) = \frac{|b - \eta^T x_o|}{n^*(\gamma\eta)} \tag{2.52}$$

where

$$\gamma = (b - \eta^T x_o)/|b - \eta^T x_o| = \text{sgn}\,(b - \eta^T x_o). \tag{2.53}$$

Proof: We assume first that $b-\eta^T x_o \geq 0$. Define

$$\beta = \max_y \{\eta^T y | n(y-x_o) \leq d_o\} \tag{2.54}$$

Then

$$\beta = \eta^T x_o + \max_{y-x_o} \{\eta^T (y-x_o) | n\left((y-x_o)/d_o\right) \leq 1\} \tag{2.55}$$

$$= \eta^T x_o + d_o n^*(\eta)$$

or

$$d_o = \frac{(\beta - \eta^T x_o)}{n^*(\eta)} \tag{2.56}$$

It is, therefore, sufficient to show that $\beta = b$.

Assume that $\beta > b$. Let y^* denote the maximizer of Eq. (2.54). Choose $0 < \lambda < 1$ such that $\eta^T(\lambda x_0 + (1-\lambda)y^*) = b$. Then

$$n(\lambda x_o + (1-\lambda)y^* - x_o) = (1-\lambda)n(y^* - x_o) < d_o \tag{2.57}$$

since $\lambda x_0 + (1-\lambda)y^* \in \pi$, this contradicts the fact that $d_0 = \min \{n(y - x_0), y \in \pi\}$. Conversely, assume that $\beta <$ b. Let y^* be such that $n(y^* - x_9) = d_0, y^* \in \pi$. Then $\eta^T y^* = b > \beta$ and $n(y^* - x_0) \leq d_0$, contradicting the fact that β is the maximum such value, thus $\beta = b$.

Next assume that $b - \eta^T x_0 < 0$. Replace b by γb and n by $\gamma\eta$ and note that π is unchanged by this. Using the first half of this proof we obtain

$$d_o = \frac{(\gamma b - \gamma\eta^T x_o)}{n^*(\gamma v)} = \frac{|b - \eta^T x_o|}{n^*(\gamma\eta)} \tag{2.58}$$

2.2.2. Generalized Simplicial Approximation

Let us define the norm body

$$B(x_o,r) \equiv \{x | n(x-x_o) \leq r\} \tag{2.59}$$

By equating this norm body with the set contained by the level contour of a joint probability density function for the statistically varying parameters, we once again approximate the yield maximization problem by the problem of inscribing the maximal norm body inside the feasible region:[14]

$$\begin{aligned} &\underset{x_o,r}{\text{maximize}}\ r \\ &\text{subject to the constraints} \\ &B(x_o,r) \subset R. \end{aligned} \tag{2.60}$$

As we have done earlier, we transform this problem to a computationally tractable problem by employing the simplicial approximation to R:

$$R = \{x | \eta_i^T x \leq b_i, \quad i=1,2,\cdots,n_f\} \tag{2.61}$$

Specifically, we have a linear programming problem of the form

$$\begin{aligned} &\underset{x_o,r}{\text{maximize}}\ r \\ &\text{subject to the constraints} \\ &d_n(x_o,\pi_j) \geq r, \quad j=1,2,\cdots,n_f, x_o \in R \end{aligned} \tag{2.62}$$

where π_j is the jth hyperplane of the simplicial approximation defined by

---

[14] It may be that as r changes, we would want to change the norm used to approximate the level contours of the pdf. This may have to be done if the level contours are radically different for different level values. However, there are many jpdf's (e.g., normal or uniform) where different level contours are similar and associated with the same norm. Since design centering is only an approximation to the yield maximization problem, we need only require that the largest norm body imbedded approximates a level contour of the jpdf within some allowed error.

$$\eta_j^T x = b_j \tag{2.63}$$

Clearly, if $d_n(s,\pi_j) \geq r$, and $x_0 \in$ R, then $B(x_0,r) \subseteq R$. Using Theorem 2 this can be written as the linear program

$$\begin{aligned} &\underset{x_o,r}{\text{maximize}}\ r \\ &\text{subject to the constraints} \\ &\eta_j^T x_o + rn^*(\eta_j) \leq b_j, \quad j=1,2,\cdots,n_f \end{aligned} \tag{2.64}$$

Observe that if $n(x) = ||x||_2$, so that $n^*(x) = ||x||_2$, we are inscribing an n-sphere in R and Eq. (2.64) reduces to the linear program derived in Section 2.1.2. Similarly, if $n(x) = ||x||\infty$, so that $n^*(x) = ||x||_1$, we are inscribing n-cubes in R.

After solution of the linear program Eq. (2.64), we can further refine the approximating polytope R as was done earlier. Specifically, begin by determining which of the n_f faces of R touch the inscribed norm body. This information is obtained by examining the slack variables associated with the linear programming solution procedure. Those constraints for which the slack variables are zero correspond to the touching faces. Of the faces which touch we choose the largest (in volume) and search in an outward normal direction from this face to find a new point on ∂R. The approximating polytope R is then inflated to include this new point as a vertex. The above procedure can then be repeated to determine a new maximum yield point.

To this juncture we have assumed that the statistical parameters have equal variances or spreads. In the event that the variances or limits of the statistical parameter distributions vary in a given problem, an additional step must be taken in the aforementioned procedures. This step amounts to a scaling of the n-dimensional parameter space with an $n \times n$ nonsingular scaling matrix T. Equivalently, one can scale instead the norm itself, as may be seen from the following:

Corollary

Let x be interior to the convex hull Eq. (2.61) and let $n(x)$ and $n_T(x)$ be norms satisfying $n_T(x) = n(T^{-1}x)$. Then

$$n_T^*(x) = n^*(T^T x) \tag{2.65}$$

and the distance in n_T norm from point x to hyperplane $\pi = \{x | \eta^T x = b\}$ is

$$d_{n_T}(x,\pi) = \frac{|b - \eta^T x|}{n^*(\gamma T^T \eta)} \tag{2.66}$$

where $\gamma \equiv \mathrm{sgn}(b - \eta^T x)$.

The proof of this corollary is similar to the proof of Theorem 2 and is omitted.

Thus a yield maximization problem with differing limits or variances of the statistical parameters may be brought into the framework developed above by defining a scaling matrix T and using the corresponding scaled norm. More specifically we define the norm body:

$$B_T(x_o, r) \equiv \{x | n_T(x - x_o) \leq r\} \tag{2.67}$$

so that with scaling the yield maximization problem is

$$\begin{aligned} &\underset{x_o, r}{\text{maximize}}\ r \\ &\text{subject to the constraints} \\ &B_T(x_o, r) \subseteq R \end{aligned} \tag{2.68}$$

The corresponding linear program which results from using the simplicial approximation $\hat{R}$ for R is

$$\begin{aligned} &\underset{x_o, r}{\text{maximize}}\ r \\ &\text{subject to the constraints} \\ &d_{n_T}(x_o, \pi_j) \geq r, \quad j = 1, 2, \cdots, n_f, x_o \in R \end{aligned} \tag{2.69}$$

or, from the above corollary

$$\begin{aligned}&\underset{x_o,r}{\text{maximize}}\, r \\ &\text{subject to the constraints} \\ &\eta_j^T x_o + rn^*(T^T\eta_j) \leq b_j, \quad j=1,2,\cdots,n_f\end{aligned} \tag{2.70}$$

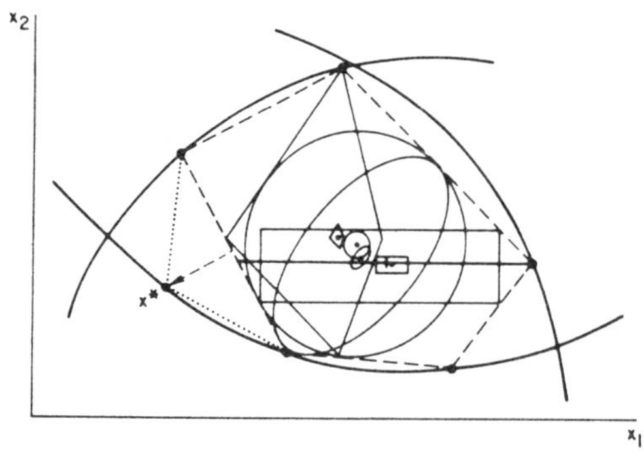

Figure 2-7: Examples of maximum norm body inscription in a polygon. The dotted line shows the effect of a typical inflation step, which expands the polygon to include the point x as a vertex.

Fig. 2-7 illustrates the result of inscribing several different types of norm bodies in the simplicial approximation. Observe that if an inflation step were to be taken the approximating polygon would expand as shown by the dotted lines. Note that asymmetrical, polygonal, and/or covariant (i.e. rotated by a nondiagonal scaling matrix T) norm bodies are easily handled by the present theory, as are degenerate (not of full dimensionality) norm bodies such as the horizontal line in Fig. 2-7.

2.2.3. Mixed Worst-Case Yield Maximization

An important situation that arises in many types of designs is one in which some of the parameters are statistical in nature with given joint jpdf's and others are worst-case parameters , i.e., the circuit must work for all values of the worst-case parameters which lie within some interval. A typical problem of this type is a circuit which is required to operate for all values of some power supply voltage within a specified range and over some range of temperatures. Thus power supply voltages and temperature are worst case parameters. In this section we describe a procedure for handling this situation.

Assume that of the n design parameters, n_w are worst case and n_s are statistical. It is convenient to partition the parameter vector x as follows:

$$x = \begin{bmatrix} w \\ s \end{bmatrix} \tag{2.71}$$

where w is the n_w-vector of worst-case parameters and s is the n_s-vector of statistical parameters. We can thus view the n-dimensional parameter space X as the Cartesian product of the n-dimensional worst case parameter space W and the n_s-dimensional statistical parameter space S. The parameters w are subject to worst-case interval restrictions which, in general, can be expressed as

$$w \in W_I \subset W. \tag{2.72}$$

Typically, W_I is an orthotope (i.e., a tolerance box), in which case

$$W_I = W_B \equiv \{w | w^l \leq w \leq w^u\}. \tag{2.73}$$

A given circuit is acceptable if the constraints on circuit performance Eq. (2.3) are satisfied, i.e.,

$$\Phi_i(w,s) \leq 0, \quad i=1,2, \cdots, n_c \tag{2.74}$$

for all $w \in W_I$. Thus the feasible region is

$$R = \{w,s | \Phi_i(w,s) \leq 0, i=1,2, \cdots, n_c; \quad \text{for all } w \subset W_1\} \tag{2.75}$$

An alternate statement of acceptable performance is obtained by defining the functions

$$\Phi_i(s) \equiv \max_{W \in W_I} \Phi_i(w,s), \qquad i=1,2, \cdots, n_c. \tag{2.76}$$

Thus acceptable performance is given by all s for which

$$\Phi_i(s) \leq 0, \qquad i=1,2, \cdots, n_c \tag{2.77}$$

and the feasible region can be described in the space S by

$$R_S \equiv \{s | \Phi_i(s) \leq 0, \qquad i=1,2, \cdots, n_c\}. \tag{2.78}$$

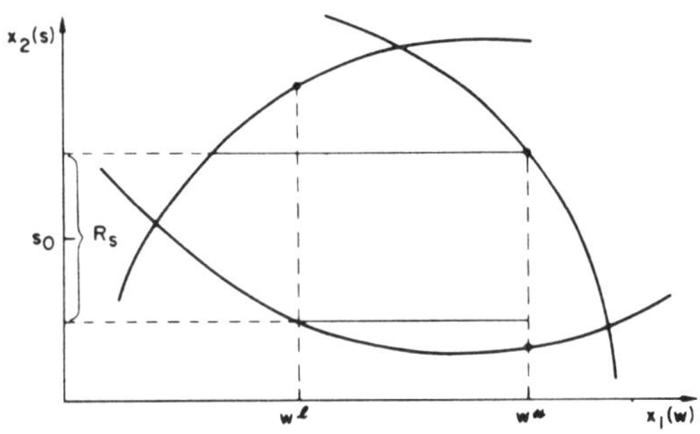

Figure 2-8: A mixed worst-case yield maximization problem with solution s_o, the midpoint of the interval R_X. R_S is the feasible region projected onto the subspace of statistical parameters.

Fig. 2-8 illustrates R_s for the hypothetical 2-dimensional case of Fig. 2-7. Given this definition of the feasible region we could employ the procedure of Section 2.1.3 to find a simplicial approximation R_S to R_S and use the procedure of Section 2.2.2 to obtain a maximum yield point. However, this approach is burdened by the fact that each step in the procedure requires the solution of the worst-case problem Eq. (2.76). Thus an alternate approach is needed. Towards this end we state the following.

Theorem 3

Assume that the region of acceptability R as given by Eq. (2.75) is convex and that the region

$$R \equiv \{(w,s) | \eta_{jw}^T w + \eta_{js}^T s \leq b_j, \quad j=1,2, \cdots, n_f\} \tag{2.79}$$

is an interior simplicial approximation to R, i.e.,

$$R \subseteq R. \tag{2.80}$$

(Note that the outward pointing normals, η_j have been partitioned as $\eta_j^T = (\eta_{jW}^T, \eta_{js}^T))$. Let Φ(s) be defined by Eq. (2.76). Then

$$R_S \equiv \{s | \eta_{js}^T s \leq b_j - \eta_{jw}^T w_o - n_W^*(\eta_{jw})\} \tag{2.81}$$

is an interior simplicial approximation to R_s as defined by Eq. (2.78) where w_0 is any point *interior* to W and

$$n_W(x) \equiv \min \{\alpha | \alpha > 0, \ \frac{x}{\alpha} + w_o \in W\} \tag{2.82}$$

Proof: We need to show that $R_s \subset$ R$_s$. Suppose that $s^* \in R_s$ so that

$$\eta_{js}^T s^* \ \leq \ b_j - \eta_{jw}^T \ w_o \ - \ n_W^* \ (\eta_{jw}), \qquad j = 1,2, \cdots, n_f \tag{2.83}$$

or

$$\frac{b_j - \eta_{js}^T s^* - \eta_{jw}^T w_o}{n_W^*(\eta_{jw})} \geq 1, \qquad j = 1,2, \cdots, n_f \tag{2.84}$$

The left-hand side of Eq. (2.84) can be interpreted as the distance of the point w_0 from a hyperplane $\pi_j(s^*)$ measured in the n_w norm where

$$\pi_j(s^*) = \{w | \eta_{jw}^T w = b_j - \eta_{js}^T s^*\} \tag{2.85}$$

which is the jth hyperplane

$$\eta_j^T x = b_j \tag{2.86}$$

restricted to the subspace $s = s^*$. Therefore, Eq. (2.84) is equivalent to

$$d_W(w_o, \pi_j(s^*)) \geq 1 \tag{2.87}$$

Since Eq. (2.87) holds for all j, the convex body W_I is contained in $R \cap \{x | s = s^*\}$. Thus for any $s^* \in R_s$ and $w \in W$, we have $\Phi_j(w, s^*) \leq 0, j = 1,2, \cdots, n_c$ which implies that

$$\max_{\omega \in W} \Phi(w, s^*) \leq 0 \tag{2.88}$$

or that $s^* \in R_S$.

Theorem 3 enables us to solve the mixed worst-case yield maximization problem by first determining a simplicial approximation to the feasible region in the full parameter space by the usual method, and using it to determine the simplicial approximation R_s of expression Eq. (2.81). Given R_s, the procedure of Section 2.2.2 could be used to maximize the yield. More specifically, we have the following linear program.

$$\underset{s_o, r_s}{\text{maximize}} \; r_s \tag{2.89}$$

subject to the constraints

$$\eta_{js}^T s_o + r_s n_S^*(\eta_{js}) \leq b_j - \eta_{jw}^T w_o - n_W^*(\eta_{jw}), \quad j = 1,2, \cdots, n_f$$

Note that Eq. (2.81) - Eq. (2.89) apply independently of the choice made for the S-space norm n_S or the tolerance region W_I. However, if $W_I = W_B$ (see Eq. (2.73)), and we choose:

$$w_o = \frac{1}{2}(w^l + w^u) \tag{2.90}$$

and

$$T_B = \frac{1}{2}(\operatorname{diag}(w^u - w^l))$$

then we get the special case

$$n_W(w) \equiv n_\infty(T_B^{-1} w) = \max \left(\frac{2|w_i|}{w_i^u - w_i^l} \right) \tag{2.91}$$

and from Eq. (2.65)

$$n_W^*(\eta_{jw}) \equiv n_1(T_B \eta_{jw}) = \frac{1}{2} \sum_{i=1}^{n_w} (w_i^u - w_i^l) |(\eta_{jw})_i| \tag{2.92}$$

Therefore, the linear program Eq. (2.89) becomes

$$\begin{aligned} &\underset{s_o, r_s}{\text{maximize}} \ r_s \\ &\text{subject to the constraints} \\ &\eta_{js}^T s_o + r_s n_S^*(\eta_{js}) \leq b_j - \sum_{i=1}^{n_w} \max \{(\eta_{jw})_i w_i^l, (\eta_{jw})_i w_i^u\} \end{aligned} \tag{2.93}$$

Several comments are in order. First, the solution that results from Eq. (2.89) or Eq. (2.93) is very much dependent upon the simplicial approximation R, as illustrated in Fig. 2-9. The vertical line $w = w^1$ of Fig. 2-9 establishes the lower (circled corners) boundary of the projected feasible region R_s. Thus when the simplicial approximation is inflated by the addition of the point x, R_s expands, and its center s_0 descends to a new level s_0. Thus in practice the approximation R should be updated and the process repeated as indicated earlier. Second, it is possible that the right-hand side of the inequality constraints in Eq. (2.93) can be negative, in which case r_s will be negative which indicates that Eq. (2.93) does not have a feasible solution. Such a situation can result for one of two reasons: (1) either the worst case constraints are too severe to allow for a solution, or (2) the simplicial approximation R to R is too crude and needs further refinement. These cases can be illustrated with Fig. 2-7, which was described previously. In Fig. 2-7, let us now regard the vertical dimension as the S space and the horizontal dimension as the W space. For the inscribed rectangle the radius r_s is positive, but for the inscribed straight line norm body r_s is zero. If the required worst-case interval is longer than the inscribed line in Fig. 2-7, the radius r_s would be negative. However, in this case the refinement of the simplicial approximation illustrated in Fig. 2-7 would lead to a positive radius on the next iteration.

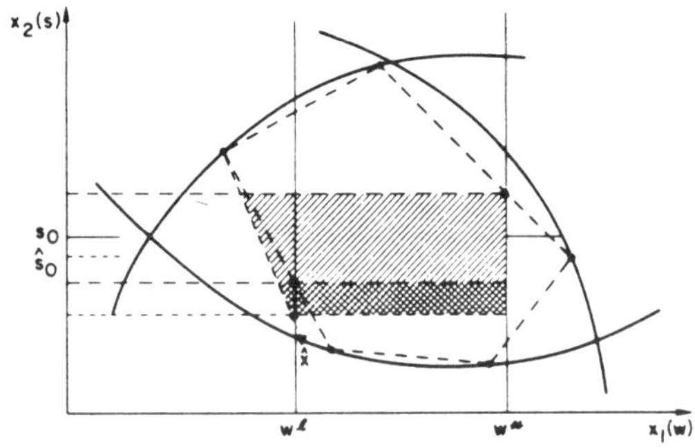

Figure 2-9: Illustration of the dependence of the solution s_o of the mixed worst-case yield maximization problem on the simplicial approximation R. As R improves as an approximation to R by the addition of the point x, s_o decreases to level s_o.

Finally, consider the special case in which $n_S = 0$ *and* $W_I = W_B$, which is the classical worst case analysis problem of determining if, for a fixed set of tolerances, whether or not the corresponding n_W-cube fits inside the feasible region. In this case the above formulation provides a significant savings in computational effort. Whereas the usual procedure would require testing for feasibility of each of the 2^{nw} vertices of the n_W-cube, all we need to do is determine whether or not the right hand side of the inequality constraints of Eq. (2.93) are positive.

2.2.4. Tolerance Assignment

The basic idea behind the yield maximization procedure described above was to inscribe a maximal norm body inside the feasible region. The norm body had a fixed aspect ratio, defined by the scaling matrix T (see Section 2.2.2).

We can view the tolerance assignment problem as a generalization of the yield maximization problem in which the aspect ratio of the inscribed norm body is allowed to vary so as to minimize an auxiliary cost function $C(t)$ [48, 89] . To see this, let $t^T = (t_1, t_2, \cdots t_n)$ be the vector of component tolerances, $T =$ diag(t) a diagonal matrix, and assume that $C(t)$ is a decreasing function of t so that $C(\alpha t) < C(t)$ for $\alpha > 1$. Then the solution of the problem

$$\begin{aligned} &\underset{x_o, t}{\text{maximize}}\ C(t) \\ &\text{subject to the constraints} \\ &(B_T(x_o, 1) \equiv \{x | n(T^{-1}(x - x_o)) \leq 1\}) \subset R \end{aligned} \tag{2.94}$$

approaches the solution of the tolerance assignment problem as R approaches R. Note that as t_i increases, the allowable deviation of x_i about its nominal increases and the cost of manufacturing component i decreases. Furthermore, observe that if all the t_i are to be identical, say, equal to r, then minimizing $C(t)$ is the same as maximizing r and this problem is the same as the problem considered in Section 2.2.2.

In general Eq. (2.94) is a nonlinear constrained optimization problem, and could be solved by use of appropriate algorithms. However, in this section we only consider the special case of inscribing an orthotope, i.e., a tolerance box, and show how, using a piecewise linear approximation to $C(t)$, Eq. (2.94) can be solved using linear programming.

It is a natural extension of the ideas of simplicial approximation to approximate the cost function $C(t)$ by a piecewise linear function. For this we assume $C(t)$ is convex and consider the convex region in the $(n + 1)$-dimensional (y,t)-space, $\{(y,t) | y \geq C(t)\}$. We then approximate this region using the ideas of simplicial approximation by

$$\left\{ (y,t) | A^T \binom{y}{t} \leq d \right\} \tag{2.95}$$

i.e., we determine a set of points $\{(y^j, t^j)\}$ such that $y^j = C(t^j)$ and then using a convex hull algorithm to find the hyperplanes which bound the convex hull of these points (the matrix A contains the hyperplane normals).

The piecewise linear approximation to $C(t)$, say $C(t)$, is then given by

$$C(t)=\min\left\{(y,t)\middle|A^T\binom{y}{t}\leq d\right\} \tag{2.96}$$

Thus Eq. (2.94) is approximated by

$$\begin{aligned}&\underset{x_o,t}{\text{minimize}}\ \mathrm{C(t)}\\&\text{subject to the constraints}\\&B_T(x_o,1)\subset R.\end{aligned} \tag{2.97}$$

Using Eq. (2.96), we can rewrite Eq. (2.97) as

$$\begin{aligned}&\underset{x_o,y,t}{\text{minimize}}\ \mathrm{Y}\\&\text{subject to the constraints}\\&A^T\binom{y}{t}\leq d\\&d_{n_t}(x_o,\pi_j)\geq 1,\qquad x_o\in R.\end{aligned} \tag{2.98}$$

Now we use the Corollary and Theorem 2 to rewrite the constraints as

$$\begin{aligned}&A^T\binom{y}{t}\leq d\\&b_j-\eta_j^T x_o\geq n_T^*(\eta_j),\qquad j=1,\cdots n_f\end{aligned} \tag{2.99}$$

or

$$\begin{aligned}&A^T\binom{y}{t}\leq d\\&\eta_j^T x_o+n^*(T^T\eta_j)\leq b_j,\qquad j=1,\cdots,n_f\end{aligned} \tag{2.100}$$

Since $n_T(x) = \max_i \{|t_i^{-1} x_i|\}$ and $t_i > 0$, then Eq. (2.97) is equivalent to (using the duality of the 1-norm and the ∞-norm)

$$
\begin{aligned}
&\underset{x_o, y, t}{\text{minimize}}\ y \\
&\text{subject to the constraints} \\
&\eta_j^T x_o + \sum t_i |\eta_{ji}| \le b_j, \quad j=1, \cdots, n_f
\end{aligned}
\tag{2.101}
$$

which we recognize as a *linear* program.

2.3. Example of Worst Case Design

We demonstrate the mixed worst-case design procedure by applying it to the CSEF "current source driver" circuit of Fig. 2-10 . The design objective is to make the reference voltage V_{REF} minimally sensitive to statistical fluctuations of the designable parameters. Here the "statistical" parameters are the resistor parameters PR1, PR2, and R3, and the transistor process parameters PIDBG, PRB, PCC. The worst-case parameters are the temperature (TC), and the power supply voltages E1 and E2. The design specifications are

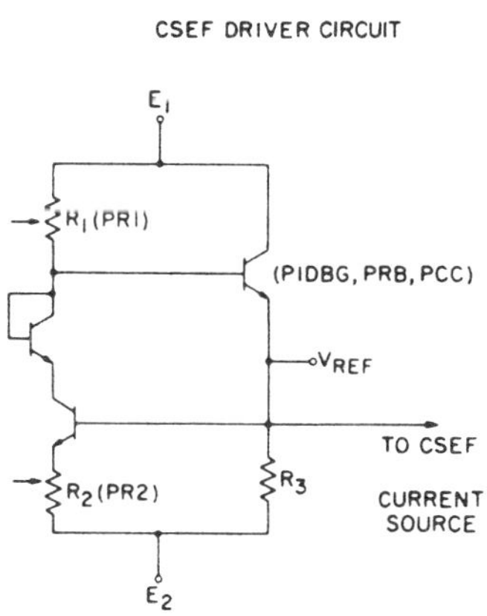

Figure 2-10: A driver circuit for a CSEF logic gate.

$$1.30 = V_{DN} \leq V_{REF} \leq V_{UP} = 1.35 \tag{2.102}$$

for *any* E1, E2, TC satisfying

$$\begin{aligned} 1.175 \leq E1 \leq 1.250 \\ 2.910 \leq E2 \leq 3.090 \\ 50 \leq TC \leq 100. \end{aligned} \tag{2.103}$$

That is, we specify that 100-percent yield is required, and seek the nominal values of R3, PR1, PR2, PIDBG, PRB, PCC which meet this requirement while maximizing the allowable tolerances on these parameters. In this maximization, we assume as before that all tolerances grow proportionately, that is, the aspect ratio of the tolerance cube is constant. We assume that all the design parameters have independent, truncated distributions, but make no assumption regarding the shape of the distributions.

The procedure is initiated by performing line searches in the positive and negative coordinate directions, yielding $2n$ points on the boundary, ∂R, of the feasible region. These points are used to obtain an initial simplicial approximation which is an n-diamond truncated by the worst case specification planes Eq. (2.103). The dual norms of the face normals of this approximation are then computed and the linear program Eq. (2.93) solved, with the scaling matrix T set to the appropriate identity matrix. However, a negative radius is thus obtained.

The procedure now undergoes a worst-case space expansion phase of the type suggested by the straight-line inscription of Fig. 2-7. However, after 20 expansions, the radius, r_s was still negative. We investigated this situation graphically, by plotting (using the multivariate piecewise linear interpolation results of [40]) various 2-dimensional cross sections of the approximating polygon. This investigation led to the suspicion that the worst-case temperature specification was the difficulty. Fig. 2-11 shows the intersection of the polygon of the simplicial approximation with a plane parallel to the PR1 and TC axes which passes through the nominal design. Inside this intersecting plane a rectangle is shown, whose height is determined by a plus and minus 100-percent tolerance interval on statistical parameter PR1. The intersection of this rectangle with the simplicial approximation shows clearly the impossibility of obtaining 100-percent tolerance on PR1. A closer look at the rightmost and leftmost points of the approximation also shows the unlikelihood of any nonzero tolerance (note the fair degree of refinement at the edges of the simplicial approximation).

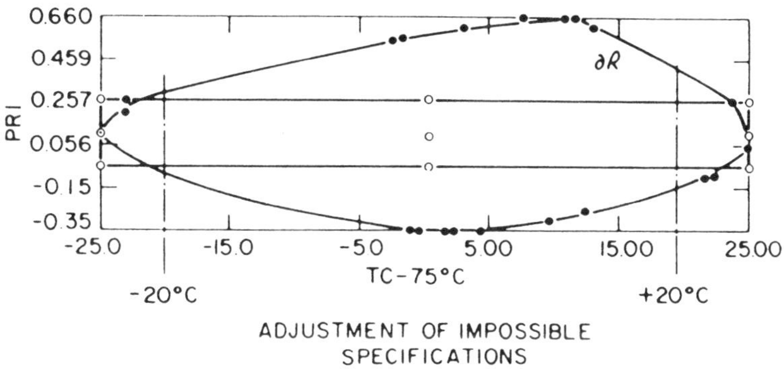

Figure 2-11: PR1-TC cross section of the simplicial approximation (after 20 negative-radius expansion steps).

This analysis suggested relaxation of the worst case temperature specification. We found that a relaxation of 5 degrees on either side of the worst case interval was sufficient to yield an acceptable solution of the problem (cf. the ±20 degree lines of Fig. 2-11).

To illustrate that temperature was the main difficulty in this design problem, we show in Fig. 2-12 the intersection of the PR1-E1 plane with the initial (i.e., $2n$-point) simplicial approximation. Note that if these were the only two dimensions of the design problem, an initial value of about 0.6 might be expected for r_s.

2.4. A Dimension Reduction Procedure[15]

The basic simplicial approximation method described so far can become computationally expensive as the dimension of the parameter space increases. In this section we introduce a scheme that may be employed to reduce the number of dimensions that need to be considered when employing the simplicial approximation method. In what follows we assume that the

[15]This section is a modified version of the paper "Dimension Reduction Procedure for the Simplicial Approximation Approach to Design Centering" by W. Maly and S.W. Director, IEE Proc. Vol 127, Pt. G, No. 6, pp. 255-259, December 1980. .

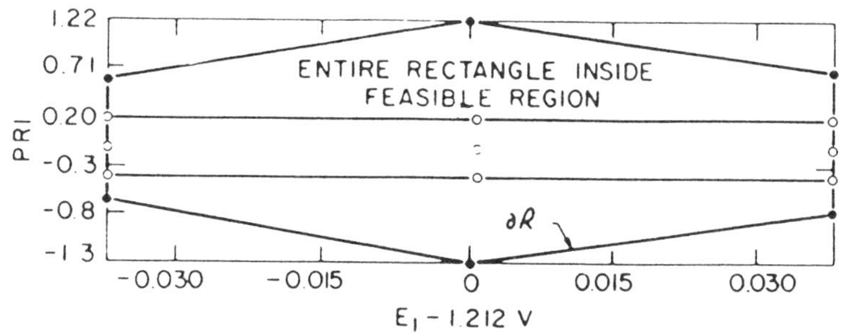

Figure 2-12: PR1-E1 cross section of the simplicial approximation (after initialization).

n-vector x denoting our design variables are normally distributed. It is possible to find [52] a linear transformation A such that:

$$x = Az \tag{2.104}$$

where z is a set of zero mean, normally distributed, random variables with unit standard deviation and correlation matrix equal to the identity matrix. This transformation, which is an orthonormalization of the random variable x, is a typical procedure applied in the method known as principal component analysis as described in [52] . Note that since A is a linear transformation, we can solve the yield maximization problem in terms of z. Furthermore, the components of z do not represent circuit parameters but are statistically independent.

One can show [52] that the contribution of each component of z to the variability of x is determined by the eigenvalues of the covariance matrix of the covariance matrix Σ_x. For strongly correlated components of x, which is the usual case for monolithic IC's (e.g. [26]), a majority of the eigenvalues of Σ_x are small. Hence, only a few of the components of z affect the spread of x and thereby influence circuit yield. We are therefore led to the idea of determining which components of z are most important in establishing the statistical behavior of the circuit and ignoring those components of z which do not affect the yield. However, fluctuations in circuit performances are dependent on both the variability of x and the sensitivity of the circuit. Therefore, the choice of the significant components of z cannot be determined using the eigenvalues of Σ_x. However, the number of the significant components of z, should be close to the number of large valued eigenvalues of Σ_x.

2.4.1. Design Centering in a Reduced Space

Consider the yield maximization problem in the space Z:

$$\max_{\mu_z} \int_{R_Z} f_z(z,\mu_z,\Sigma_z)dz \tag{2.105}$$

where μ_Z Σ_Z and R_Z are mean, covariance matrix and the region of acceptability in the transformed space Z: $Z=\{z=A^{-1}x|x\in X\}$ respectively. The difficulty with this problem is that R_Z is only know implicitly in terms of z. In the simplicial approximation method , a sequence of polyhedral approximations $(H_1,H_2 \cdots ,H)$ to R_Z such that $H_1 \subset H_2 \subset \ldots \subset H$, is generated. The yield maximization problem Eq. (2.105) is then approximated by the problem

$$Y_{max} = Y_H(\mu_Z{}^{opt}) = \max_{\mu_z} \int_H f_Z(z,\mu_Z,\Sigma_z)dz \tag{2.106}$$

Finally, the problem Eq. (2.106) is replaced by the problem of inscribing the largest hypersphere (which is a level contour of the j.p.d.f $f_z(z)$) inside H. Observe that in the simplicial approximation method, the approximation H is generated independently of the level contours of $f(z)$. (Note that the level contour associated with a joint normal distribution is defined by

$$l(r,\mu_z)=\{z|n_2(z-\mu_z)=r\} \tag{2.107}$$

where n_2 is the Euclidean norm). By taking into account the radius of the level contour and the location of R_Z, it is possible to determine if there are dimensions of the space defined by z that have little or no effect on the yield. Elimination of these superfluous dimensions allows us to perform design centering in a reduced dimensional space at a reduced cost. We will show later that for the case of IC's it is usually possible to significantly reduce the dimension of the parameter space.

We now describe the dimension reduction heuristic. Let H_0 denote the initial polyhedral approximation to the feasible region which contains the origin of Z. The vertices of H_0 are the points of intersection of R_Z with the positive and negative co-ordinate axes, and can be found by means of one-dimensional line searches. Let these vertices be denoted by z_k, for $k = \pm 1, \pm 2\ \ldots, \pm n$. Observe that if

$$\|z_{\pm j}\| > r \tag{2.108}$$

for some $j = 1,2, \ldots, n$, then the level contour $l(r, \mu_Z)$ is completely contained within H_0 along the jth co-ordinate axis. Furthermore, if $r > 3$ (which is three times the standard deviation in the normalized space) and Eq. (2.107) holds, then reasonable deviations in the jth component of Z are likely to have little effect on the yield. Thus, we are motivated to project H_0 onto the $n-1$ dimensional subspace ${}^jZ$ obtained from Z when the jth dimension is dropped. We denote this projection by ${}^jH$ and solve the problem:

$$y^j_{max} = \max_{\mu_z} \int_{jH} {}^jf_z({}^jz, {}^j\mu_z, {}^j\textstyle\sum_z) d^jz \tag{2.109}$$

where ${}^jz$ and $f_Z({}^jz, {}^j\mu_Z, {}^j\Sigma_Z)$ are equivalents of z and j.p.d.f. in subspace ${}^jZ$. Because of the discussion above, we expect that $Y^j_{max} \simeq Y_{max}$.

With the above as background we can formalize the heuristic as follows:

Step 1

Determine the initial polyhedron H_0 in space Z using searches along the co-ordinate axis of Z to find the $2n$ vertices, $z_k, k = \pm 1, \pm 2, \ldots, \pm n$.

Step 2

Determine for each j (see the next Section) if

$$\|z_j\| \geq \alpha \quad \text{for} \quad j = \pm\, 1,2, \cdots, \pm\, n \tag{2.110}$$

where $\alpha > 3$ is an arbitrarily assumed value. Drop from Z all dimensions j for which Eq. (2.110) holds.

Step 3

Continue with the original simplicial approximation procedure in the reduced space Z'.

Note that the space reduction decision is made based on the initial approximation of $R_{Z'}$, i.e. H_o. The polyhedron H_o does not carry information about the entire region R_Z. Thus, there exists the possibility that use of this algorithm would result in the dropping of some directions of Z which might result in significant yield improvement. However, note that if α is assumed to be large enough, then the sensitivity of the yield with regard to the variability of Z along the dropped dimension is small. Therefore, it is unlikely that the solution of the yield maximization problem of expr. Eq. (2.105) lies outside of space Z'.

2.4.2. Discussion

We have described above an algorithm for solving the parameter space dimension reduction problem . The proposed method is especially useful for monolithic ICs composed of a large number of elements which are strongly correlated. Application of the proposed algorithm presented in [62] indicates that the dimensions of the parameter space can be reduced significantly and the dimension of the reduced space is associated with the number of significant eigenvalues of Σ_X.

We expect these results to hold in general because, for monolithic technologies, the correlations between circuit elements are determined for the most part by the properties of the manufacturing process and should be independent of the structure of the circuit and the number of circuit elements. Thus we expect that, even for a circuit composed of several dozen elements, the dimension of the yield maximization parameter space would be of the order of that shown in the example. Hence, solving the design centering problem in the reduced space Z' instead of the ordinary parameter space X, the simplicial approximation method can successfully be applied to a much larger class of ICs. However, note that the benefits of the proposed heuristics are dependent on such factors as the shape of H_o and the value of α. By decreasing α, we increase the probability that space Z' does not include the solution of the yield maximization problem, Eq. (2.105). Hence, the value of α should be chosen carefully. We emphasize again that the proposed method was intended as an extension of the simplicial approximation method. However some of the features of this method may be useful in other approaches to the statistical design of monolithic IC's.

2.5. Fabrication Based Statistical Design of Monolithic IC's[16]

Consider again the yield maximization problem:

$$\max_{\mu_z} \int_{R_X} f_z(z, \mu_x, \Sigma_x) dz \tag{2.111}$$

where R_X is the region of acceptability.

To this juncture we have not been very specific about the proper choice of designable parameters. In particular, the vector x has been used to denote any of the circuit parameters that are randomly varying due to disturbances in the manufacturing process, such as resistances and capacitances, threshold voltages, transistor β's. However, the simplicial approximation method that has been developed to solve the yield maximization problem is based upon the assumptions that

A1: It is possible to obtain any desired value of μ_x by means of adjusting the process parameters and/or the layout of the IC.

A2: The covariance matrix, Σ_x is independent of the circuit and the process parameters.

Unfortunately, these assumptions are not necessarily valid for integrated circuit applications. In particular:

1. The mean values of some "designable parameters" are not

[16] This section is based on the paper "Fabrication Based Statistical Design of Monolithic IC's", by W. Maly, A.J. Strojwas and S.W. Director, " Proceedings of the 1981 International Symposium on Circuits and Systems, pp 135-138, Nov., 1981

designable in the sense that they are not necessarily adjustable to any solution of Eq. (2.111). For example, the mean value of the threshold voltage of an MOS transistor cannot be adjusted to any arbitrary value.

2. In general, the mean values of some "designable parameters" are not independent of one another. For example, the mean values of the β's of two n-p-n integrated transistors are strongly correlated to one another and cannot therefore be set to arbitrarily different values.

3. The covariance matrix may not be independent of the mean values. For example, the mean and variance of the resistance of a monolithic resistor are dependent on its geometry.

We show below that by judicious choice of truly independent designable parameters, and an alternate formalization of the yield maximization problem, we can develop a method which is more suitable for the design of monolithic IC's. A key step in the solution of the yield maximization problem will be the ability to simulate manufacturing process.

2.5.1. Independently Designable Parameters in Yield Maximization of Monolithic IC's

We wish to determine under what conditions the solution of the yield maximization problem Eq. (2.111) is technologically realizable. Assume that the manufacturing process and the performance of a circuit can be controlled by a set of *primary design variables* denoted by the vector z which are deterministic, physically independent and controllable quantities (e.g. temperatures, diffusion times, mask dimensions, etc.). Observe that changes in the values of the primary design variables cause the means of the random variable x to change, i.e., μ_x is a function of z. If the solution of Eq. (2.111) is μ_x^{opt} then it is technologically realizable if there exists a z^{opt} such that

$$\mu_x^{\text{opt}} = \mu_x(z^{\text{opt}}) \tag{2.112}$$

It will always be possible to find a z^{opt} such that Eq. (2.112) holds if the components of the vector $\mu_x = \{\mu_{x_1}, \mu_{x_2}, ...\}$ are independent and strictly monothonic functions of components of the vector $z = \{z_1, z_2, ...,\}$. We call any random variable x_i, whose mean, μ_x is a strictly monothonic function of a primary design variable z, a *designable random variable.* If the value of μ_x can be arbitrarily changed while keeping the other components of μ_x constant, then x_i is said to be an *independently designable random variable.* Hence, a solution of Eq. (2.111) will be technologically realizable if every component of x is an independently designable random variable.

We now wish to determine which parameters are the primary design variables and which parameters are independently designable random variables. In order to motivate this discussion consider the following simple example. Let w and l be the width and length of a rectangular window in a photolithographic mask used to fabricate an integrated resistor. Due to imperfections in the photolithographic process the dimensions of this window in SiO_2 are described by random variables L and W, respectively. Assume for simplicity that $W = w + \Delta F$ and $L = l + \Delta F$, where ΔF is a zero mean normally distributed random variable with standard deviation $\sigma_{\Delta F}$. One can show that if both l and w are much greater than $3\sigma_{\Delta F}$ then, $F = L/W$ is a normally distributed random variable with mean $m_F = l/w$ and variance

$$\sigma_F^2 = [(w - l)/w^2]\, \sigma_{\Delta F}^2 \tag{2.113}$$

Furthermore, it can be shown that the actual resistance of the integrated resistor is also normally distributed random variable R such that $R = F{*}R_s$ with mean μ_R and standard deviation σ_R^2 given by

$$\mu_R = \frac{1}{W}\mu_{R_S} \tag{2.114}$$

$$\sigma_R^2 = \mu_{R_S}^2 [(w - l)/w^2]\, \sigma_{\Delta F}^2 + \left(\frac{1}{W}\right)^2 \sigma_{R_S}^2 \tag{2.115}$$

where R_s is a normally distributed random variables with mean μ_{R_s} and standard deviation σ_{R_s} which represents the sheet resistance. Thus R is a *designable random variable* because any specific mean of its resistance can be obtained by adjusting the ratio l/w. Moreover, it is an *independently designable random variable* because the mean values of the resistance of the different resistors in the circuit can be chosen independently of one another. Hence, it would appear that the nominal resistance values of integrated resistors could be used as optimization variables for solving the yield maximization problem Eq. (2.111). However, since μ_R and σ_R are dependent on each other this choice of variables would violate assumption A2.

Note that this observation can be extended to all electrical parameters of monolithic elements because both the nominal values and higher order moments of these electrical parameters are related to the mean and variances of the set of mask dimensions, as well as their ratios or window areas. (The moments of dimension ratios or areas are dependent of each other (e.g. see Eq. (2.115))). Hence none of electrical parameters of monolithic elements should be considered as a designable random variable.

For the examples discussed above, we can choose as optimization variables the means of L and W. (L and W are independently designable random variables because $\mu_L = l$ and $\mu_W = w$ and their variances are independent of μ_L and μ_W, respectively). In general, since any IC design can be described in terms of the mask dimensions, then the mean of the random variable $z = \mu_z + \Delta F$, representing the dimensions of the IC elements, which are related to the mask dimensions z and imperfections in the photolithographic process ΔF, can be chosen to be an optimization variable in the yield maximization problem . (Note that z is a primary design variable as before). Finally, note that if among the components of primary design variables are parameters controlling the manufacturing process such as times, temperature, etc., then the random variable Z should also include the random variables representing these quantities (e.g., if t represents the temperature of the oxidation process in the description of the process then Z should be extended by the component $T = t + \Delta T$ where ΔT is the inaccuracy in the process temperature).

2.5.2. Process Simulation and Yield Maximization

Since the components of x cannot represent the electrical parameters of the IC, the yield maximization problem as formulated in Eq. (2.111) needs to be modified. Towards this end let x be as before, a vector of parameters which characterize the electrical parameters of the network. Let z be a vector of independently designable random variables representing the mask dimensions of the elements in the IC and the basic process parameters. The circuit equations Eq. (2.1) can therefore be written as:

$$g(\xi, \dot{\xi}, x, t) = 0, \qquad 0 \leq t \leq t_f \tag{2.116}$$

$$C(x, t) = 0 \tag{2.117}$$

where Eq. (2.117) represents a model of the manufacturing process. Using Eq. (2.116) and Eq. (2.117) one can define the feasible region R_Z in the space Z , $(z \in Z)$. We can now formally state the yield maximization problem as

$$Y_{max} = \max_{\mu_z} \int_{R_Z} f_z(z, \mu_z, \Sigma_z) dz \tag{2.118}$$

where $f_z(z, \mu_z, \Sigma_z)$, μ_z, Σ_z are the jpdf, means and covariance matrix of Z, respectively. Thus by adding to the previous statement of the yield maximization problem the constraints describing the manufacturing process and properly choosing elements of z, the yield maximization problem for monolithic IC's can be solved by the methods proposed in this chapter.

Observe that, in general, the dimension of z is much larger than the dimension of x. (For instance, the number of variables describing the layout coordinates of the zigzag resistor is much larger than the number of its electrical parameters). Thus, the computational expense of approximating the feasible region R_Z in the space of independently designable parameters Z could be much larger than the computational expense of approximating R_X in the space of circuit electrical parameters X. Fortunately, we may employ a technique based upon process simulation to alter the statement of the yield maximization problem so as to reduce the computational effort.

Assume that we have an approximation, $R_{X'}$, to the feasible region R_X. Since the random variable x is dependent on the primary design variable z, at least some moments of x are also dependent on z. Let $g[m]_x^1(z)$ and $g[S]_x^1(z)$ denote those moments of X which are dependent on z, and let $g[m]_x^2$ and Σ_x^2 denote those moments which are independent of z. The yield maximization problem can then be stated in the following way:

$$Y_{max} = \max_{\mu_z} \int_{R_X} f_x(x, \mu_x^1, \Sigma_x^1, \mu_x^2, \Sigma_x^2) dz \qquad (2.119)$$

Note that any solution of Eq. (2.119) is technologically realizable and because we are in a lower dimensional space, the cost of the solution of Eq. (2.119) should be less than cost of the solution of Eq. (2.118). Of course a key step in being able to solve Eq. (2.119) is ability to simulate the manufacturing process.

Let us now consider the process simulation technique we need for solving Eq. (2.119). Because variations in the process parameters, as well as circuit elements, are due to the unavoidable disturbances in the manufacturing process, the random variable x is dependent on both the primary design variable z and the process disturbances.

$$x = P(z, D) \qquad (2.120)$$

where P ( ) is a model of the manufacturing process relating the electrical circuit parameters x to z and D. Observe that the disturbances of a process are associated with a particular manufacturing facility but are independent of the particular circuit being designed and fabricated. Thus, the jpdf describing D, $f_D(d)$, needs to be identified only once for each process. By employing Eq. (2.120) we can estimate moments of x in terms of z and therefore solve Eq. (2.119). In the next chapter we describe a process simulator that has the capabilities we need.

Chapter 3
Statistical Process Simulation

3.1. Introduction

In the last chapter we indicated that yield maximization methods, which take into account inherent fluctuations in the IC manufacturing process, must be supported by statistical process and device simulation. It will also be demonstrated in the following chapters of this book that such simulation is a necessary ingredient not only for circuit design but also for process and device development, as well as IC process diagnosis and control. This chapter describes the statistical process and device simulator FABRICS, which can mimic the stochastic nature of the manufacturing process. This simulator is based on concepts first proposed in [59], and then in [61, 65] and [83]. FABRICS consists of two major parts: a process simulator and a device simulator. The models implemented in FABRICS allow for the statistical simulation of typical semiconductor devices manufactured by a variety of fabrication processes.

In this chapter we introduce the basic concept of statistical process simulation and outline specific features and requirements that should be met by such a simulation. Then the current structure of FABRICS is discussed in some detail. As will be seen, FABRICS can successfully mimic the fluctuations of any IC fab line due in part to the ability to "tune" it to the behavior of a given line. Such tuning involves determining the statistical moments (e.g., means and standard deviations) of the jpdf's that characterize the process disturbances. Hence we describe the program PROMETHEUS [105] that has been developed for this purpose. Finally in the last section of this chapter we describe the current implementation of the entire FABRICS system.

3.2. Statistical Process Simulation

The goal of statistical process simulation is the emulation of the IC manufacturing process. Specifically we are attempting to mimic the relationship between process control , and IC layout as well as process disturbances and the performance of fabricated IC's. Therefore, statistical simulation must involve modeling of the process disturbances, modeling of the processing steps, modeling of IC devices and circuit simulation. Conceptually, therefore statistical process simulation should have a structure as depicted in Fig. 3-1.

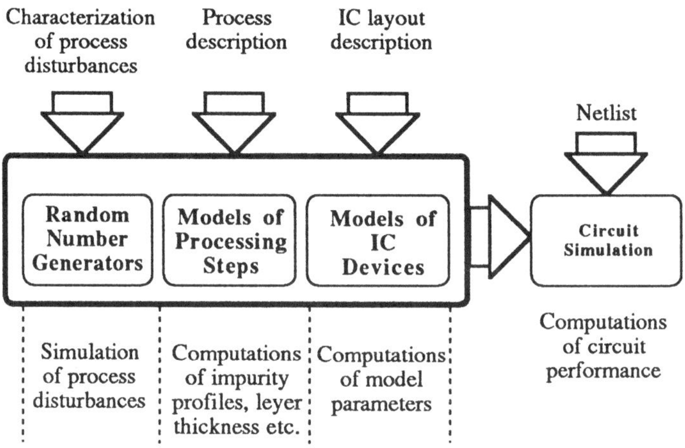

Figure 3-1: Steps needed for statistical process simulation.

In this section we discuss in more detail the general philosophy of statistically based process simulation, and techniques for modelling disturbances, processing steps, and devices.

3.2.1. Methodology[1]

Recall the model of the IC manufacturing process introduced in Chapter 1. In this model the manufacturing process is viewed as a sequence of

[1] This section is based upon the paper "Statistical Simulation of IC Manufacturing Process" by W. Maly and A.J. Strojwas, IEEE Trans. on CAD, July 1982, Vol. CAD-1, No. 3, pp. 120-131.

manufacturing operations (see Fig. 1-2) with each manufacturing operation consisting of a processing step usually followed by an inspection and selection step. The outcome of each process step depends upon a set of control parameters, a set of disturbances, and the outcome of previous process steps.

In order to formally describe the simulation procedure we need to define a number of quantities. Let X, as before, be a vector of random variables that describe parameters of the IC elements or values of the in-line measurements. The vector C represents the process control parameters, e.g., temperatures, times, gas fluxes, etc., while the vector L represents the layout parameters (e.g. mask dimensions.) Finally, let D be a vector of random variables representing disturbances in the process.

Now the jth process step can be characterized by a model of the form:

$$X_j = g_j(X_{\alpha 1}, \ldots, X_{\alpha k}, L_j, D_j, C_j) \text{ for } j = 1 \cdots, m \tag{3.1}$$

where

X_j - the jth component of X, is a vector of *physical parameters* that describes the outcome of a fabrication step (e.g., oxide thickness, parameters of impurity distribution, misalignment, etc.),

$X_{\alpha i}$ - is an αi-th component of the vector of physical parameters obtained from the models of previous fabrication steps (therefore αi must be less than j),

L_j, D_j, C_j - are the vectors containing those components of L,C, and D, that affect the jth manufacturing step.

Models of circuit elements can be assumed to be of the form:

$$X_j = h_j(X_{\beta 1}, \ldots\ldots, X_{\beta k}) \; for \; j = m+1, \cdots, n \tag{3.2}$$

where $X_{\beta i}$ are process characteristics that affect j-th electrical parameter of the fabricated device.

Assume that models of the form described above are available. Then, conceptually, simulation of the process can be carried out in the same manner as the process itself. In other words the process simulator can have a structure (i.e. a computational flow) that is identical to the physical flow of the actual manufacturing process. For instance, computations needed to determine characteristics of the N-channel MOS process may be viewed as the sequence of procedures, which describe processing steps depicted Fig. 1-1 in Chapter 1, executed in appropriate order.

Hence in general, because we desire a simulator that can handle various processes and a rather large number of devices, the actual simulator can be implemented as a library of process step models that are executed in the appropriate sequence.

A key element of the simulation strategy discussed in this chapter is the choice of models that have been symbolically described above by the functions g_i and h_i. While it is possible to develop a detailed and accurate mathematical characterization for each step in the manufacturing process, and for each device being manufactured, the use of such models in a statistically based simulator would be computationally prohibitive. The reason for this is that such models often require the solution of a system of nonlinear partial differential equations. Since statistical simulation requires large sample sizes, and hence repeated solutions of these models, the computational cost is high. What we need then is a set of models that are orders of magnitude faster to evaluate while at the same time able to provide sufficient accuracy. Therefore, we propose to use analytical models (or in some cases efficient numerical models) of each manufacturing step and circuit element that explicitly characterizes the dependence of the output parameters on the input parameters. (Thus, the functions g_j and h_i in Eq. (3.1) and Eq. (3.2) can be thought of as approximations to the solutions of the partial or ordinary differential equations in the accurate numerical models.)

A major concern of statistical process simulation is the accuracy that can be achieved during simulation. In statistical simulation, accuracy can be measured in terms of the difference between parameters of the probability distribution of X generated by the statistical process simulator and estimated from the measurements. Note that the process control C and layout L are known and given. Thus in order to achieve good agreement between the simulated and measured data, parameters of the probability distribution of process disturbances have to be determined. This can be accomplished by employing statistical extraction methods on both the data generated by the models and the data collected from in-line and test pattern measurements. If the models of the process steps and circuit elements are complex enough to account for the most significant physical phenomena, good accuracy can be obtained. In addition, if the models are physically meaningful, statistical simulation can be accurate enough within the neighborhood of a design to provide predictive capabilities.

Hence we feel that in order to effectively perform statistical simulation of the IC manufacturing process it is necessary to:

- use, as often as possible, analytical models of processing steps and IC elements;
- use models whose parameters may easily be employed to mimic process disturbances;
- use models whose parameters can easily be identified;
- build simple, but physical, models to provide not only identification flexibility and good simulation performance, but also physical fidelity.

To achieve such a goal it is necessary to very carefully make a decision about the choice of the simulation algorithms applied to model process disturbances, processing steps and elements of IC devices.

3.2.2. Modeling of Process Disturbances[2]

In Chapter 1 we described a process disturbance as a randomly changing environmental factor that affects performance of the fabricated IC. Usually such factors are complicated functions of both time and spacial coordinates, and in the general case must be defined as stochastic processes. In practice, however, a process disturbance can be accounted for by treating one or more of the parameters of the element and/or process model as random variables.

To see this, consider the fluctuations of an MOS transistor threshold voltage. This voltage varies among transistors within an IC chip because of

[2] This section is based upon the paper "Modeling of Random Phenomena in CAD of IC - A Basic Theoretical Consideration", by W. Maly, Proceedings of ISCAS, pp. 427-430, 1985.

instabilities inherent in the process that affects the gate oxide thickness, substrate doping, channel width and length etc. The actual value of each of these quantities is determined by a number of stochastic processes (including randomly fluctuating temperatures, water resistivities, etc.). While stochastic processes such as these are the "true" disturbances in the manufacturing process, they are difficult and impractical to model for two reasons. First, the exact relationships between these environmental factors and quantities of interest, such as the threshold voltage, are not well understood. Second, we are not really concerned with the time dependences of these relationships. Actually what we are interested in is the final value of the oxide thickness, densities of various charges in the gate oxide etc. Therefore, it is more convenient to treat disturbances as random variables in the model of threshold voltage. Such variables implicitly imitate the influence of the original process disturbances provided that they cause the threshold voltage to be distributed in a manner similar to what can be observed in an actual manufacturing process.

It is important in the above approach to recognize that in order to properly model the actual process fluctuations, the random variables chosen as process disturbances must have a special structure. Consider again the MOS transistor threshold voltage and note that this voltage can vary due to among other reasons, variations in gate oxide thickness. Usually the gate oxide thickness variations observed within a single IC chip are small because all transistors from a single IC chip are located very close one to another and therefore have very similar " *process histories* " . This holds true for the majority of the parameters affecting threshold voltage. Thus we can expect the threshold voltages of all transistors within a single chip to be similar and close to a certain mean value. (See Fig. 3-2 a and c.) Such is not the case, however, if we compare transistors from two different chips. Since the mean temperature for one chip can be quite different than the mean temperature for another chip, the gate oxidation process will be different for different chips, especially if they are on two different wafers. (See Fig. 3-2 b .)

The above discussion suggests that each process disturbance should have a mean value characterizing "average" process conditions for some local area. The actual conditions for a specific location within this area can be modeled by a variable randomly fluctuating about this "local mean value" with some "local standard deviation". These local parameters should also randomly change from one local area to another (See Fig. 3-2 d.).

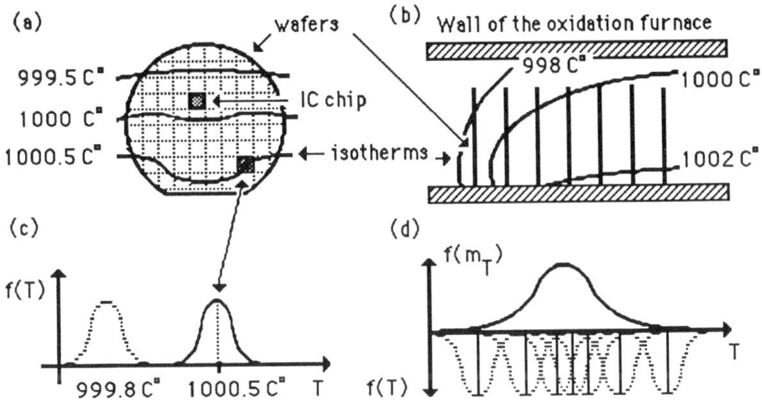

Figure 3-2: Distribution of the temperature in the gate oxidation process on the wafer (a) and in the furnace (b). Probability density function associated with the temperature in a chip (c) and means of the chip temperatures in the furnace (d).

To account for local (intra-die) and global (inter-die) variations in device parameters, we can employ a *multilevel structure* [65] for the random variables that characterize process disturbances. With such a structure, disturbances are generated by hierarchically defined random number generators (RNGs) at levels that correspond to natural divisions, i.e., at the lot, wafer, chip and device levels. These process disturbances must also be chosen such that they are:

- statistically independent;
- independent of particular features of IC layout, such that the simulator can be used for different ICs manufactured in the same fabrication process;
- independent of process controlling parameters.

We further assume that the process disturbances have either normal or log-normal distributions. Note that such an assumption is not only convenient but also realistic. Variations of model parameters are usually caused usually by a large number of independent random variables. Consequently it is highly likely that the distributions of random variables representing disturbances are normal. Note that to completely specify these disturbances we must identify their means and standard deviations. Since the parameters that characterize the process disturbances cannot be obtained by direct measurements, they have to be identified through in-line and test structure measurements. (This identification problem can be formulated as a nonlinear optimization problem [60]. But it is a highly dimensionality and efficient decomposition techniques have to be applied. A system for the automatic extraction of the probability distribution of process disturbances, called PROMETHEUS [104, 105] described in Section 3.3.)

The above requirements can be met through the use of a *two-level random variable* which is a special case of a *mixture* [30]. Specifically, it is assumed that the two-level random variable X is described by a *lower level* probability density function $f_l(x,M_1,M_2,....,M_r)$ and a set of *upper level* probability density functions $f_{uj}\ (m_j\ ,m_{j1}\ ,m_{j2}\ ,\ .....\ ,m_{jk}\ )$, for $j=1, 2,. .... , r$. Moments of the lower level density function M_i , $i=1, 2\ , ...\ ,r$, are random variables and their distributions are defined by the upper level density functions.

To describe a single process disturbance a number of two-level random variables are used and are arranged in a hierarchical structure containing several levels of *locality*. On the lowest level of locality, a two-level random variable is used to describe fluctuations in a very small area (usually within a single chip). Moments of this random variable are described by set of two-level random variables describing a higher level of locality, say a wafer. Moments of random variables characterizing this level of locality can again be defined as two-level random variables on yet a higher level of locality, say a lot. (And so on as illustrated in Fig. 3-3.) This scheme, shown in Fig. 3-4, is implemented in FABRICS.

The theory of mixtures [30] provides the basic tools for the analysis of two-level random variables. In particular, this theory is useful for determining the appropriate distribution for functions of the upper level random variables to ensure certain properties for the distribution function of the lower level two-level random variable. Of special importance is that, in general, a two-level random variable is not Gaussian. Since experimental data, as well as, the Central Limit Theorem [22], suggest that process disturbances are Gaussian, we must define the two-level random variable in a manner assuring its normal distribution.

Fabrication	$f_f(m_l, m_{f1}, m_{f2}, \ldots\ldots, m_{fp}) \Rightarrow \mathbb{M}_{\mathbb{f}}$
Lot	$f_l(m_w, \mathbb{M}_{\mathbb{f}}) \Rightarrow \mathbb{M}_{\mathbb{w}}$
Wafer	$f_w(m_{ch}, \mathbb{M}_{\mathbb{w}}) \Rightarrow \mathbb{M}_{\mathbb{ch}}$
Chip	$f_{ch}(x, \mathbb{M}_{\mathbb{ch}}) \Rightarrow \mathbb{X}$

Figure 3-3: Hierarchical structure of process disturbances.

The normal distribution of a two-level random variable can be assured in several ways. The simplest way is to assume that both lower and upper distribution functions are Gaussian and that the mean value of the lower level density function is a random variable while other moments are deterministic variables. In other words, it is convenient to assume that the two-level random variable X is defined by a lower distribution, $N_l(x, M, \sigma)$, where $N()$ denotes the Gaussian probability density function, and σ is the deterministic value of standard deviation. This case M is a normally distributed random variable defined by the upper distribution function, $N_u\ (m, m_m, \sigma_m)$, where m_m and σ_m denotes the mean value (of mean values) and standard deviation (of mean values), respectively. Below we will show that the two-level random variable defined in this way has a Gaussian probability distribution and and that moments of such a variable can be easily determined by moments of upper and lower density functions.

To compute the probability density function of the two-level random variable $X, f_X(x)$, we assume that B denotes the event that $X \in [x, x+\Delta x]$, and A_i denotes the event that $M \in [m_i, m_i + \Delta m]$. From the law for the total probability [22] we have

$$P(B) = \sum_{i=-I}^{i=I} P(B/A_i)p(A_i) \tag{3.3}$$

where $P(B)$ denotes the probability of the event B, and

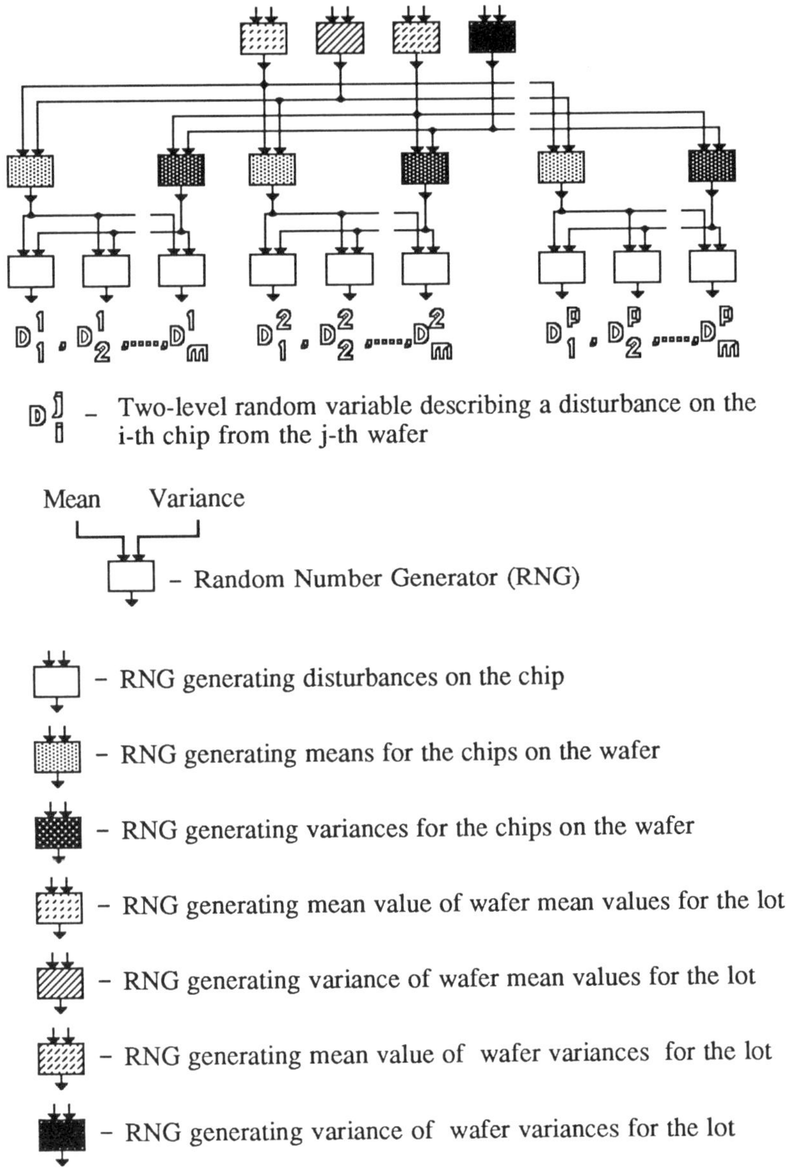

Figure 3-4: Generation of process disturbances.

$$P(B/A_i)=\int_{x}^{x+\Delta x} N_l(x,m_i,\sigma)dx \tag{3.4}$$

$$P(A_i)=\int_{m_i}^{m_i+\Delta m} N_u(m,m_m,\sigma_m)dm, \tag{3.5}$$

Δx and Δm are small, and $m_i=(I-i)\Delta m$ for $i=-I,-I+1,...,I$. Now assuming that $\Delta x\rightarrow 0, \Delta m\rightarrow 0$ and $I\rightarrow\infty$ one can show that Eq. (3.3) takes the form:

$$f_X(x)=\int_{-\infty}^{\infty} N_l(x,m,\sigma)N_u(m,m_m,\sigma_m)dm \tag{3.6}$$

Then integrating Eq. (3.6) with explicit expressions describing the Gaussian distribution one can show that

$$f_X(x)=\frac{1}{\sqrt{2\pi(\sigma_m^2+\sigma^2)}}\exp\left(-\frac{(x-m_m)^2}{2(\sigma_m^2+\sigma^2)}\right) \tag{3.7}$$

Hence X has an ordinary Gaussian distribution with mean equal to m_m and standard deviation equal to the sum of the local variance σ^2 and variance of mean values σ_m^2 .

Note now that the two-level random variable can be used to define a *two-level Gaussian random vector* . It is defined as a set of two-level normally distributed random variables with mean values which are given by a normally distributed random variable common for all components of this vector. (See Fig. 3-5) and the same lower level variance σ.

It is obvious that components of the two-level random vector are normally distributed random variables, with means and standard deviations defined by Eq. (3.7). The only nontrivial feature of this vector is that its components are correlated. Here we will show that the covariance matrix of this vector can be easily expressed in terms of appropriate variances of lower and upper density functions.

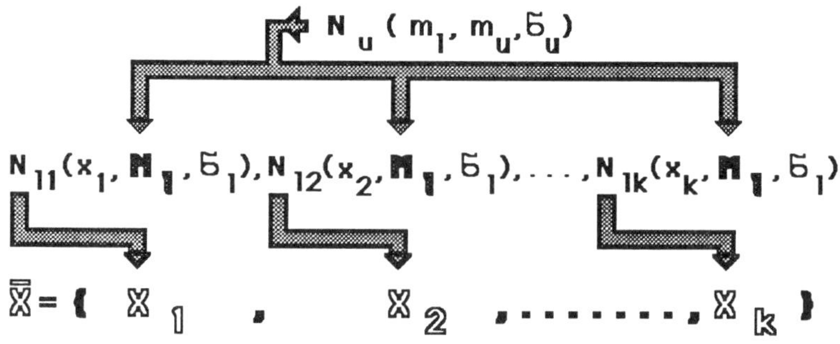

Figure 3-5: Two-level Gaussian random vector.

To compute the correlation factor between the two components of the two-level random vector note that any two-level random variable defined above can be viewed as a sum of two independent random variables, one representing the local variations of the two-level variable and the other one describing fluctuations of the mean values, i.e. representing the global portion of variability of this variable. This is evident by recognizing that Eq. (3.7) is a distribution function of a sum of two independent random variables. Thus by analyzing the correlation factor between the i-th and j-th component of the two-level Gaussian vector, X_i and X_j, respectively, we can assume that $X_i = Z_i + Z_0$ and $X_j = Z_j + Z_0$, where Z_i and Z_j and Z_0 are independent random variables, Z_i and Z_j have the same Gaussian distribution, $N(z,0,\sigma)$ and Z_0 is distributed with the density function $N(z, m_m, \sigma_m)$. From the above one can compute the correlation factor:

$$\rho(X_i, X_j) = \sigma_m^2 / (\sigma_m^2 + \sigma^2) \tag{3.8}$$

Hence, the components of the two-level Gaussian random vector are characterized by a mean of m_m, a standard deviation of $\sqrt{\sigma_m^2 + \sigma^2}$ and correlation factor given by Eq. (3.8).

The above discussion implies some important consequences as far as statistical simulation of the IC manufacturing process is concerned. Two of the most important of these are:

- By using two (or more) level random variables one can appropriately model disturbances in the process because in this way it is easy to distinguish local and global components of the disturbance. Such a distinction is absolutely necessary for the accurate modeling of correlations between IC parameters. (See Eq. (3.8)).

- Assuming a structure for the disturbance such as shown in Fig. 3-4, one can significantly simplify the process identification procedure.

Therefore for statistical simulation as described in this book, multilevel random variables are used. To provide the desired level of flexibility it is assumed that all moments are random variables, as well. (In reality, however, a majority of disturbances are modeled as normally distributed random variables with normally distributed mean values.)

3.2.3. Process and Device Models for Statistical Simulation

Traditional approaches to process simulation, such as that implemented in SUPREM [2], employ *numerical models* to characterize each fabrication process step. These models are usually expressed in terms of partial differential equations (e.g. the diffusion equation) that are solved using numerical techniques to produce the nominal profiles of impurities in silicon. The impurity profiles can then be used by device simulators such as SEDAN [21] and MINIMOS [100], in which semiconductor device models are described by a system of partial differential equations, (e.g. Poisson, transport and continuity equations) that are solved numerically to produce device parameters. While such simulators can produce results that are accurate for the deterministic case, as mentioned above, they are prohibitively expensive when used for statistical investigations.

To alleviate these difficulties, FABRICS employs analytical models [61, 65] which are solutions of the partial differential equations that describe each fabrication step under a set of restricted or simplifying conditions that have been found to yield reasonable results. (In fact analytical modeling techniques have recently become more popular in the area of process and device simulation. Several programs have been developed in which either analytical models or a combination of numerical and analytical models are implemented (for example, the two-dimensional process simulator, SUPRA [51], and the process and device simulator, PRIDE [88]). These approaches offer a reasonable compromise between accuracy and simulation efficiency.

A detailed description of fabrication step and device models implemented in FABRICS is beyond the scope of this book and can be found in [82]. However, in order to illustrate the modeling approach employed by FABRICS we will present a brief description of the models implemented for ion implantation followed by thermal redistribution and some elements of the MOS transistor model.

3.2.4. Models of Ion Implantation

The profile of impurities after ion implantation can be taken from approximate solutions to a rigorous theory for implantation proposed by Linhard, Scharff and Schiott [56]. These approximate solutions may be tabulated [36] [14], and then used as a basis for a model for statistical simulation. In particular, the two-dimensional profile of impurities after implantation may be described by two one-dimensional profiles (vertical and lateral) such that:

$$N(x,y)=N_1(x)N_2(y)$$

where $N_1(x)$ is a Gaussian profile [132], and $N_2(y)$ is a complementary error function given by the formula [1]:

$$N_1(x)=N_{max}exp\frac{-(x-R_p)^2}{2\sigma_p^{\,2}}$$

$$N_2(y)=N_1(0)erfc\frac{-y}{\sqrt{2}\,\sigma_l}$$

where

$$N_{max}=\frac{Q}{\sqrt{\pi/2}\,\sigma_p}\left(1+erf\frac{R_p}{\sqrt{2}\,\sigma_p}\right)$$

Note that $N_1(x)$ is the profile in the vertical direction, $N_2(y)$ is the profile in the lateral direction, and R_p,σ_p and σ_l are the range, straggle and lateral straggle, obtained by least-squares fits of the tabulated data as functions of implantation energy [36]. The straggles, σ_p and σ_l are disturbed by a *straggle variation* $\Delta\sigma$ to account for process disturbances in this step.

Thermal processes following an implantation affect the impurity profile. Two models may be used to model this effect. In cases where the thermal step results in a diffusion length $\sqrt{Dt}$ (where D is the diffusion coefficient and t is time) that is small compared to the range of the vertical profile R_p, the step may be modeled as an anneal, otherwise, it can be modeled as an impurity redistribution.

The annealed profile retains its Gaussian distribution, with some increase in the straggle [124]. The increase in the vertical and lateral straggles may be expressed as:

$$\sigma_p = \sqrt{\sigma_p^2 + D_p t}$$

$$\sigma_l = \sqrt{\sigma_l^2 + D_l t}$$

where $D_p t$ and $D_l t$ are the squares of the lateral and vertical diffusion lengths. D_l and D_p are the lateral and vertical diffusion coefficients, and are increased to model damage-enhanced diffusion.

The vertical impurity profile after thermal redistribution can be approximated by the sum of two profiles, one resulting from the out-diffusion of impurities into the growing oxide, and the other from redistribution without oxidation [53], [130]. Boron and phosphorus distributions can be modeled by the same profile function, with different segregation coefficients, while the arsenic distribution model can take into account high doses typical of arsenic implants [29].

The analytical solution used for boron and phosphorus assumes that the profile before redistribution is Gaussian. Thus when modeling multiple thermal redistributions, the profile after each redistribution step must be approximated by an equivalent Gaussian profile. The lateral profile of redistributed impurities can be modeled by an analytical approximation to Smith's function.

The physical parameters controlling the profile models, namely diffusivity, segregation coefficient, and oxide growth rates, should be treated as disturbances.

3.2.5. MOS Transistor Model

The MOS transistor model implemented in FABRICS generates a complete set of MOS device parameters compatible with the SPICE circuit simulator [80]. The process simulator portion of FABRICS passes to the device simulator portion of FABRICS a set of *physical parameters*, that includes impurity concentrations, sheet resistances, gradient voltages and oxide thicknesses. These are used by the device simulator to generate device parameters such as:

- *Threshold Voltage*, which is corrected for short-channel and narrow width effects, as well as the non-uniform profile in the gate region;

- *Intrinsic Transconductance*, which is calculated with an electron mobility that is decreased for the surface inversion layer;

- *The dimensions of the device*, which are modified by lithographic variations such as line-width variations and misalignments, as well as by the lateral diffusion of the source and drain into the gate region;

- *Internodal and Substrate Zero-bias Capacitances*, which are calculated from the geometrical dimensions and overlaps.

Here again analytical models are used in which parameters are provided by process simulation portion and also are extracted from the layout description.

3.2.6. Structure of the Simulator

In this section we describe the structure of the most recent version of FABRICS, known as FABRICS [83]. FABRICS is composed of two major components, FAB1, the process simulator and FAB2, the device simulator.

FAB1 contains a library of models for each manufacturing step as well as libraries of functions that simulate impurity profiles and extract in-line parameters from the simulated profiles. These routines are controlled by a *process supervisor* which sequences through the manufacturing steps of a particular process, such as NMOS, and simulates profiles and extracts *physical parameters* that can be later used for device simulation by FAB2. This *process supervisor* routine is similar to a fabrication process flow chart, and can be user-defined. Thus, a *customized* version of the process simulator for a particular manufacturing process may be created. The

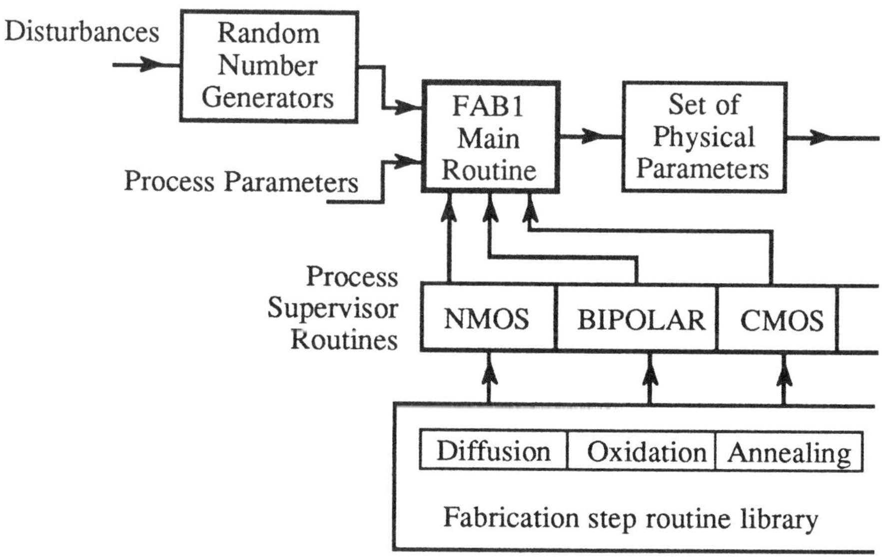

Figure 3-6: Structure of process simulator FAB1.

structure of FAB1 is shown in Fig. 3-6; the inputs consist of process parameters, process disturbances, and run control parameters. Some of the process models included in the manufacturing step library are:

- *Diffusion* of impurities into silicon is modeled as a two-step process consisting of predeposition and drive-in . Segregation of impurity atoms between silicon and silicon dioxide is taken into account. Diffusion models for the three basic dopants, i.e. phosphorus, boron and arsenic, are included.

- *Ion Implantation* profiles are determined from analytical models that account for annealing and multiple thermal redistributions of ion-implanted impurities such as described in 3.2.4.

- *Thermal oxidation.* The model for oxide growth takes into account the changes in oxidation rates due to various oxidizing ambients (steam, dry or wet oxygen), as well as for heavily doped substrates.

- *Photolithography.* The statistical model for lithographic operations accounts for linewidth inaccuracy caused by misalignment, and by fluctuations inherent in optical and etching operations.

- *Surface treatment processes* are modeled statistically by random variables representing process fluctuations (e.g. surface potential, surface recombination velocity, surface state density, and specific contact resistance).

FAB2 uses the physical parameters generated in FAB1, combined with the layout of the devices, to produce device model parameters such as those of the Shichman-Hodges model of an MOS transistor or the Gummel-Poon model of a bipolar transistor. These parameters can then be used in a circuit simulator to predict circuit performance.

The model parameters are computed using functions stored within the device model library; and a model *template* is then filled and passed to the circuit simulator. The current version of the device model library contains models for:

- MOS transistors;

- Bipolar transistors: n-p-n and p-n-p (lateral and substrate);

- MOS capacitors;

- Diffused resistors;
- Interconnections (parasitic resistances and capacitances);

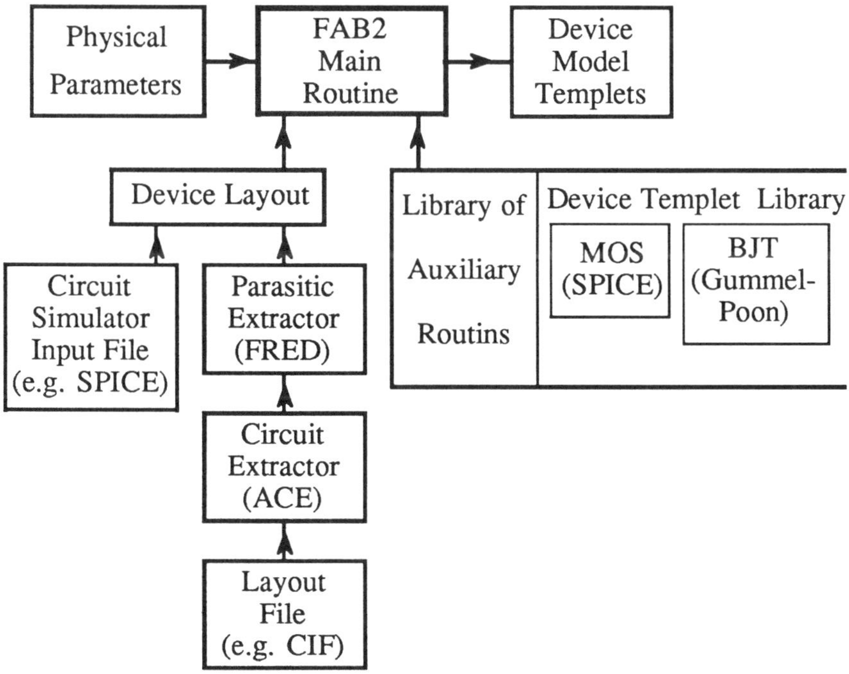

Figure 3-7: Structure of device simulator FAB2.

Recently, a new set of device modeling procedures has been developed. Detailed information about these models can be found in [82].

Fig. 3-7 shows the structure of FAB2. Information about the dimensions of the various devices can be entered to FAB2 through a typical SPICE input file containing the nominal device dimensions or through the description of the layout using Caltech Intermediate Form (CIF). In the latter case, individual device dimensions are derived from a CIF file using a circuit extractor such as ACE [39] and an interconnect extractor such as FRED [122]. This layout information is divided into two parts: device dimensions, which are used in FAB2 to calculate device model parameters, and interconnections, which produce equivalent parasitic resistances and capacitances.

3.3. Tuning of Process Simulator with PROMETHEUS[3]

As mentioned earlier, FABRICS models process disturbances through the use of a set of hierarchical random number generators. The accuracy of this approach is dependent on our ability to determine the moments of multilevel random variables that characterize these disturbances. Unfortunately disturbances are not directly measurable and very often are not easily observed. In this section we present a technique for evaluating these quantities. This technique was developed in [105] and is the most general approach to the problem previously defined and solved for the simple case in [60]. We refer to this process as *tuning*.

Simply stated tuning is the process in which moments of random number generators modeling the process disturbances are adjusted to obtain a good agreement between data measured in the process and that generated by the process simulator FABRICS (see Fig. 3-8). In the general case tuning can also be seen as a process in which the basic statistical characteristics of an IC process are extracted. Hence, the general problem to be solved is extraction of the statistical characteristics of an IC manufacturing process. Thus we begin our discussion of tuning with a discussion of the process characterization problem .

3.3.1. Mathematical Formulation

To precisely define the process characterization problem it is convenient to introduce the following notation. The n-dimensional vector $\mathbf{d}$ will be used to represent the n process attributes that characterize the process disturbances. The m-dimensional vector $\mathbf{p}$ will denote the m process attributes that describe the process conditions. Finally, the k-dimensional vector y represents the k process outputs that describe the characteristics of a manufactured circuit and its components (e.g., physical and electrical device parameters such as threshold voltages, parasitic capacitances etc).

Let us assume now that the physical phenomena that occurs during the IC

[3] This section is based upon the paper: "Parameters Extraction for Statistical Process Characterization" by C.J. Spanos and S.W. Director in IEEE Trans. on CAD, Vol. CAD-5, No. 1, pp. 66-78, 1986.

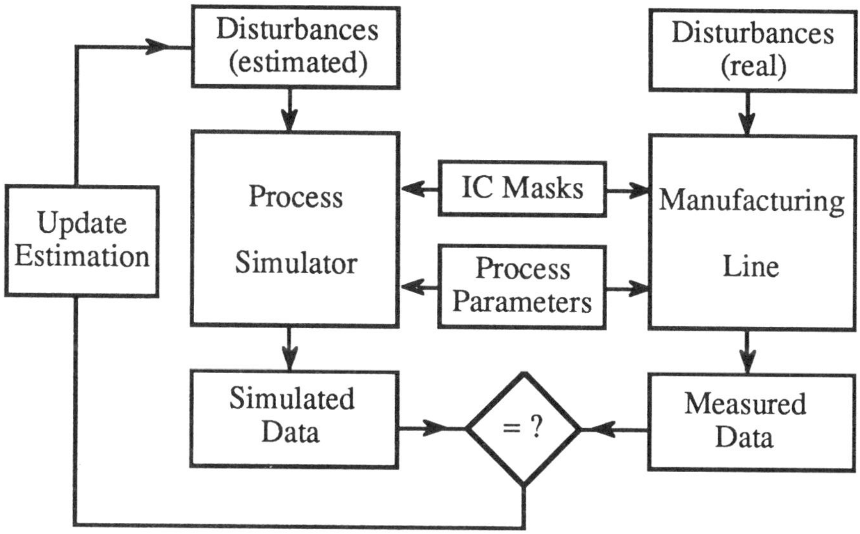

Figure 3-8: IC process characterization.

fabrication process, for given values of the process parameters, are modeled by a system of nonlinear algebraic equations:

$$\mathbf{y} = \mathbf{F}(\mathbf{d}). \tag{3.9}$$

Note that Eq. (3.9), is deterministic and therefore can only be used to describe the behavior of the process for a specific set of parameter values. In fact, by appropriate selection of values for the process disturbances, Eq. (3.9) describes the *nominal* behavior of an IC process. It is possible, however, to use Eq. (3.9) to capture the statistical attributes of an IC process if we allow the vectors **d** and **y** to be random variables of the form:

$$\mathbf{d}=\mathbf{d}_o+\mathbf{d}_e \tag{3.10}$$

$$\mathbf{y}=\mathbf{y}_o+\mathbf{y}_e. \tag{3.11}$$

Here $\mathbf{d}_o$ and $\mathbf{y}_o$ are vectors of deterministic values and represent the nominal values of the disturbances and the outputs of the process, respectively: $\mathbf{d}_e$ and $\mathbf{y}_e$ are realizations of the random processes.

More specifically d represents the N independent statistical variations that are superimposed on the nominal values of the disturbances of an IC fabrication process and $\mathbf{y}_e$ represents the K (usually correlated) statistical variations superimposed on the nominal values of the outputs of an IC process. Also, we will note by θ^d the R-dimensional vector of the significant moments that describe the statistics of $\mathbf{d}_e$ affecting the process. Similarly, we use θ^v to denote the S-dimensional vector of the moments that describe the statistics of $\mathbf{y}_e$.

At this point let us amplify what is meant by nominal values. While observing an actual process, the engineer must decide on what constitutes nominal values and what, therefore, constitutes deviations from nominal. While this choice is somewhat arbitrary, it is convenient to select as the nominal a value so that the *median of the statistical component is zero.* Since we are presently concerned with normal distributions, we will then choose the nominal value so that the *mean of the statistical component is zero.* Such a choice greatly simplifies the solution of our characterization problem, by allowing the use of regression models created during what we call the deterministic stage of the extraction for sensitivity analysis during the statistical part of the extraction.

Based upon our discussion of nominal values, we define the nominal values of $\mathbf{d}_o$ and $\mathbf{y}_o$ such that Eq. (3.9) is satisfied when $\mathbf{d}_e$ and $\mathbf{y}_e$ are set to zero. In other words, once $\mathbf{y}_o$ is determined from measurements, a $\mathbf{d}_o$ is chosen so that

$$\mathbf{y}_o = F(\mathbf{d}_o). \tag{3.12}$$

In this way, the statistical variation superimposed on the outcome of the process, $\mathbf{y}_e$, is the direct result of the statistical variation superimposed on the process conditions, $\mathbf{d}_e$. We also assume that the statistical distribution of each of the disturbances $\mathbf{d}_e$ is completely characterized by the respective elements of the vector of moments θ^d, and the joint statistical distribution of outputs $\mathbf{y}_c$ is completely characterized by the vector of moments, θ^v.

The problem of statistical process characterization can now be defined as follows: Given a model that describes an IC process, as well as the nominal values of the outcome of the process $\mathbf{y}_0$ and the statistics of the joint probability density function of the outcome of the process $\mathbf{y}_e$ (both of which can be inferred from measurements), identify $\mathbf{d}_o$, and the corresponding moments of the independent distributions of $\mathbf{d}_e$.

3.3.2. Methodology of Solution

As it was proposed in [105] the above problem can be effectively solved in two phases. The first involves identification of $\mathbf{d}_o$ given a particular measured value Y_o of $\mathbf{y}_o$. This operation is referred to as the *pretuning phase* and involves the solution of the problem:

$$\min_{\mathbf{d}_o} \| Y_o - F(\mathbf{d}_o) \|_2. \tag{3.13}$$

Solution of Eq. (3.13) is straightforward using standard nonlinear programming methods. Having found $\mathbf{d}_o$ the second phase, called *final characterization phase* , is carried out to identify the statistics of $\mathbf{d}_e$. Application of nonlinear programming techniques to this part of the problem is considerably more involved.

Since the process disturbances and the process outputs are explicitly related through Eq. (3.9), we may assume that an implicit relationship exists between θ^y (i.e., the statistical moments of the process outputs) and θ^d (i.e., the statistical moments of the process disturbances) as well. We denote this relationship by

$$\mathbf{G}(\theta^y, \theta^d) = 0 \tag{3.14}$$

Assuming that moments $\theta_0^{\ y}$ are given from measurements one can find moments of the disturbances by solving minimization problem:

$$\min_{\theta^d} \| \mathbf{G}(\hat{\theta}^{\,y}, \theta^d) \|_2. \tag{3.15}$$

The problem is, however, that since the relationship characterized by Eq. (3.15) is not explicitly available, nor is it easily deduced from Eq. (3.9), an approximation must be devised. Furthermore, successful solution of the minimization problem in Eq. (3.15) implies several assumptions. Namely, that the jpdf of $\mathbf{y}_e$ can indeed be described by a number of known moments, and that the jpdf of $\mathbf{d}_e$ is of known form. It also implies that the estimators for the moments to be used in the extraction are robust, so that the best is made of the limited number of measured samples that we have at our disposal, and that formal termination criteria have been developed, so that the solution will not be overly dependent on the particular subset of the population that has been measured.

Finally let us recall that the statistical IC process characterization problem, treated in its entirety, requires the simultaneous extraction of moments of multilevel random variables representing process disturbances. Because there may be as many as, for instance, forty independent process disturbances and as many as eight statistical moments per disturbance, the respective extraction problem becomes prohibitively complicated. Thus, we are motivated to decompose the hierarchical statistical characterization problem into a sequence of non-hierarchical (flat) subproblems. We will then perform the extraction on each of these subproblems and subsequently reconstruct the hierarchical description of the process disturbances.

The non-hierarchical statistical extraction subproblem is to be solved in two stages. These stages are: the *extraction stage* , where the problem is formulated as a deterministic minimization and then solved; and the verification stage where the validity of the solution found during the extraction stage is evaluated statistically. The extraction stage can be further simplified if (as already mentioned) it is done in two phases. These phases are: the *pretuning phase* where the nominal values of the disturbances are extracted Eq. (3.13), and a *final characterization phase* where the statistical moments of the disturbances are extracted by solving Eq. (3.15). In [105] the above approach is described in more detail.

3.4. The Process Engineer's Workbench[4]

The Process Engineer's Workbench (whose structure is shown in Fig. 3-9) integrates the capabilities of FABRICS, as well as a number of tools that

---

[4]This section is based upon the paper "The Process Engineer's Workbench" by A. J. Strojwas and S. W. Director, IEEE Journal of Solid-State Circuits, No.2, April, 1988, pp. 377-386.

have been developed to work in conjunction with FABRICS (such as PROMETHEUS and the statistical design rule developer STRUDEL [97]), into a unified process synthesis environment. The The original implementation of FABRICS as a stand alone simulator, in terms of program size, constitutes less than 10 percent of the entire PEW system . This environment has a flexible, graphics-based user interface that allows the designer a great deal of freedom in using the various tools. It also has the capability of interacting with other technology-oriented CAD tools such as SUPREM and PISCES .

In what follows we describe some of the features of PEW. We begin with a discussion of the features of the most current version of FABRICS that make it ideally suited to forming the heart of the Process Engineer's Workbench. We then overview the user interaction that is possible with the PEW during process synthesis and discuss some of the operational details of the system. Next we describe the internal data structures which are essential for the incremental simulation. The concept of process compilation and subsequent simulation capabilities of the customized version of the simulator within PEW is described. Finally, we present an application of PEW for synthesis of a commercial state-of-the-art fabrication process. We conclude with a brief discussion of possible extensions to PEW.

3.4.1. Process and Device Simulation

While PEW can perform all of the basic functions of FABRICS, it employs a different program structure in order to achieve greater flexibility and ultimately achieve increased functionality. In PEW (see Fig. 3-9) the process model library contains models for each of the process steps. Specifically, models are included for ion implantation, annealing, diffusion (predeposition and drive-in), multiple redistribution, oxidation (2-d LOCOS and thin oxide growth), epitaxy , all lithography steps, deposition , and dry etching . These models can be called in an arbitrary sequence so as to be able to simulate arbitrary process flows (e.g. bipolar, MOS or BICMOS). Further, simulation can be carried out in an incremental fashion, i.e. the designer can examine the results after each step of the process, or after the entire fabrication process. The complete sequence of fabrication steps associated with a given process, along with the list of in-line measurements or device parameters for which values are desired, is specified in PEW by a *process supervisor*. Processor supervisors for an arbitrary process can be easily generated by the processor

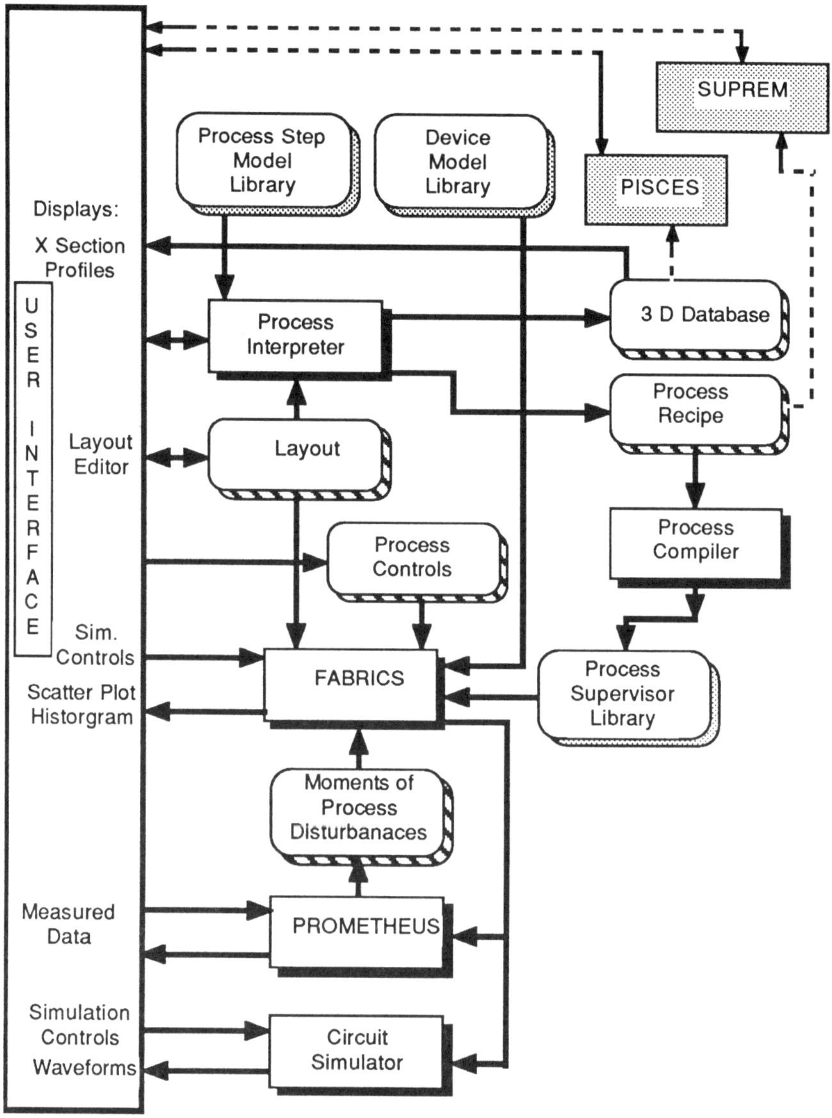

Figure 3-9: Process Engineer's Workbench.

interpreter/compiler. Once a process supervisor has been developed it is stored in the *process supervisor library*.

Models in the process model library are either analytical or efficient numerical in nature. The moments that characterize the probability distributions associated with these parameters can be extracted from in-process measurements, that have been made on physical test structures, through the use of PROMETHEUS.

The process simulator in PEW can generate two-dimensional impurity distributions and oxide profiles, as well as device topography. It can also generate device parameters such as threshold voltage as well as a number of in-process parameters, such as sheet resistances, junction depths and layer thicknesses.

The device model library stores models for each device type. Examples of device types currently stored in this library are bipolar transistors, (including vertical n-p-n, substrate and lateral p-n-p's), small geometry MOSFETs, IC diodes, Schottky diodes, arbitrary capacitors (e.g. p-n junction and overlap) and resistors (for arbitrary conducting paths). These models can be used to generate both device current-voltage characteristics and device model parameters to be used by circuit simulators.

3.4.2. User Interaction-Process Synthesis

A process recipe is entered into the PEW by means of a versatile menu- and form-driven user interface [57]. This interface contains a mask editor which allows the user to define the layout of the basic device structures. Since the process step library contains models of all the major processing steps, and is capable of generating the two-dimensional impurity distributions and oxide profiles, as well as device topography (deposition, lithography and etching models are included), full process integration capabilities can be obtained.

Each fabrication step in a process recipe is entered by making the appropriate selection from a menu in the order that is performed in the real manufacturing line. A set of process control parameters must be specified for each fabrication step. This is accomplished through the use of a form. Fig. 3-10 shows the parameter entry form for an oxidation step. Observe that each line of the form contains information about the type of parameter whose value is to be entered and the allowable range of values for that parameter. Incorrect parameter values are not accepted and the user is prompted for a correction. To provide a reasonable starting point for inexperienced users, default values of control parameters for each process

Oxidize Specifications

Name [string] pre_ox1

Type [enum] Dry

Time <0:30000> ([min:] sec) [string] 600

Temperature (K) [float] <800 : 1600> 1223

Oxygen partial pressure [float] <0 : 1> 0.45

HCL partial pressure [float] <0 : 1> 0.00

Figure 3-10: Oxidation entry form.

step are suggested.

If a process step requires a lithographic mask, a mask editor may be invoked from the menu as shown in Fig. 3-11. This figure illustrates the capabilities

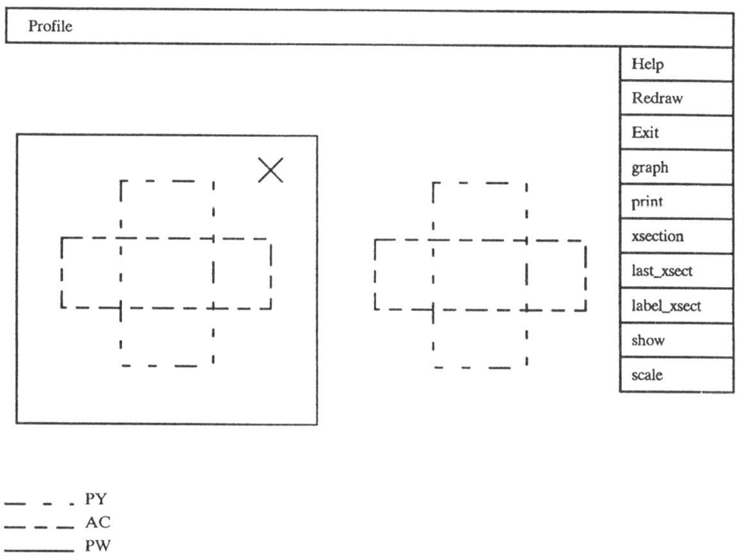

Figure 3-11: Mask editor.

of the mask editor which is functionally equivalent to a standard layout-editing system. Each masking layer has a unique code which is explained on the left-hand side of the screen.

In PEW, incremental process simulation is accomplished through the use of the *Process Interpreter* [47] coupled to a three-dimensional database. The database contains information about all of the layers that define each device structure and its electrical properties. For each step in the process, the interpreter invokes the appropriate process model routine and then modifies the 3-D database after simulating the step. The interpreter also has an extensive error checking capability and can detect such errors as improper values of process control parameters, e.g. an insufficient value for the implantation energy that may result in the entire dosage of impurities staying in the screening oxide. Errors in the process sequence, such as omission of the photoresist lift-off step, can be detected as well.

Incremental simulation is particularly useful if the designer wishes to observe the effects of process controls on the output parameters (e.g., layer thicknesses and junction depths) of an individual process step. Process controls can be manipulated until the desired values of the output parameters are obtained.[5]

To facilitate evaluation of process simulation results, a number of display capabilities have been provided in PEW. Both one-dimensional impurity profiles and two-dimensional cross-sections through the devices of interest can be displayed. Furthermore, a user can request an arbitrary cross-sectional display by specifying the cross section of interest with a line in the mask editor.

3.4.3. Internal Data Structures

In order to support the above features, a full three-dimensional representation of the chip structure is used. Specifically, the chip area is divided into *regions*, and each region is defined by a unique *stacking of layers* (e.g., p-type impurity region, oxide, polysilicon, insulator, metal). Fig. 3-12 illustrates the representation of a typical NMOS transistor, while

---

[5] In a future release of the system, an optimization feature will be included that will allow PEW to automatically determine the values of a set of process control parameters that will result in a specified set of output values.

Fig. 3-13 illustrates the representation used internally in which the boundaries between regions are approximated in the piecewise linear form.

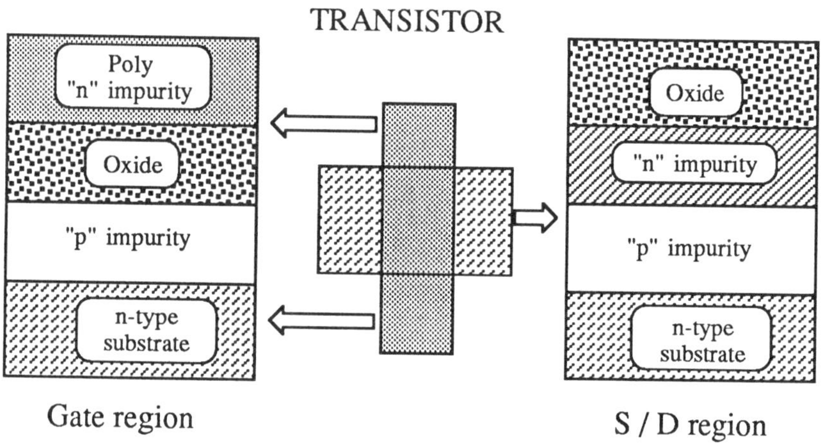

Figure 3-12: Simplified structural representation.

This data representation can be viewed as an extension of other IC representations such as the Caltech Intermediate Form (CIF). Unlike CIF, where only conducting layers are considered, the region/layer approach includes information about all layers. In addition, each layer has more detailed information concerning the material properties. For example, a diffusion layer in CIF would be represented in the PEW database using a profile function which specifies a 2D impurity concentration. This representation is updated after each new lithographic mask is entered into the system. For instance, when simulating an implantation, the interpreter checks each region in the database to determine where the implantation will penetrate. A new layer representing an impurity profile is added to the layer stack for each region in which penetration occurs. Extraction routines in the interpreter generate information about the process which can then be displayed by the user.

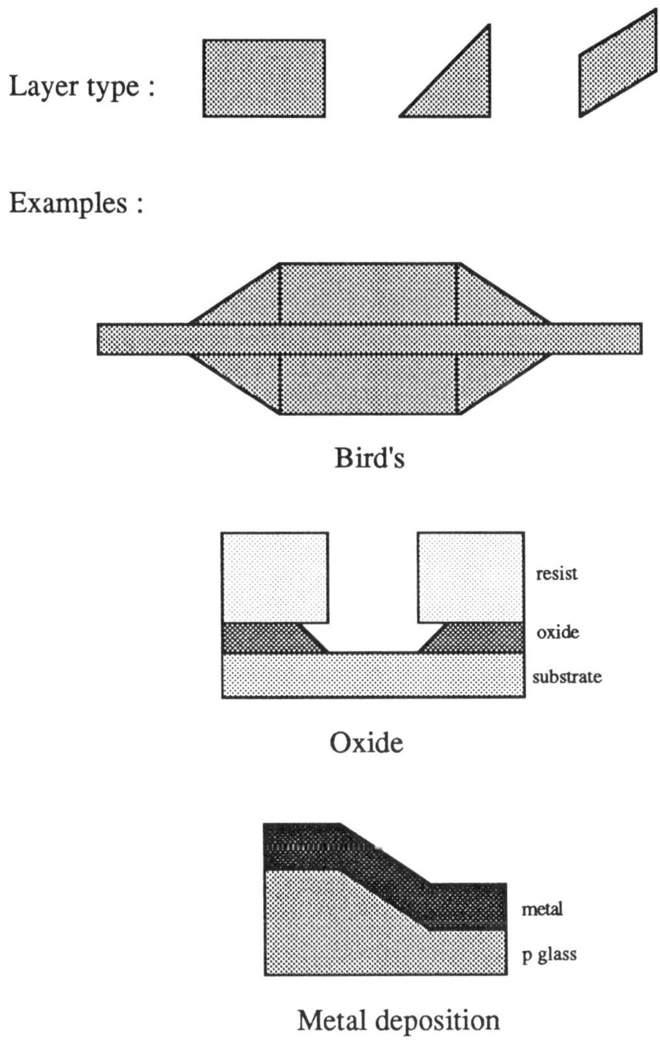

Figure 3-13: Piecewise linear structural representation.

3.4.4. User Interaction - Compiled Simulation

After the designer is satisfied with the simulation results for a complete process, the process recipe can be translated by the process compiler into a process supervisor linked with the appropriate device models from the process and device model libraries and stored in the process supervisor library. More specifically, the process compiler generates the C-code for each process step needed to simulate the effect of this step on the 3-D database. In addition, the C-code extracts physical parameters for use in the FABRICS device simulator. All the data files needed by the process supervisor are also produced by the compiler. The process supervisor and associated data files are then linked with the necessary routines in the model libraries. Linking a supervisor is a two-step process. In the first step, the process supervisor is linked to a diagnostic version which has extra checks to assure that the devices are correctly defined. Although the process information indicates that device structures are valid, extra checks are added to verify that the electrical characteristics of the devices are acceptable. Once the diagnostic version has been successfully run with the new process, the supervisor is linked to the statistical version which can be used to generate data for a large chip population. In essence then, a "customized" version of FABRICS is created and efficient process/device simulation can be now performed. Note that by changing the output generator in the Process Interpreter, an input file for SUPREM can be created. Moreover, the data structures that reside in the database at the end of the complete process simulation can be transferred to a two-dimensional device simulator such as PISCES or MEDUSA. Such capabilities are planned for the next generation of PEW.

Actual execution of the customized version of FABRICS is controlled via the user interface. In the nominal simulation mode , the output from the device simulator in FABRICS can be examined via the two-dimensional plots of electrostatic potential (as illustrated in Fig. 3-14), depletion layer spreading, and the extracted device model parameters such as threshold voltage and transconductance of MOSFET's. In the statistical mode, samples of devices with different dimensions that typically occur due to mask misalignment can be generated. The statistical distributions of the device model parameters can be analyzed by statistical post-processing routines invoked from the user interface. It is also possible to display the statistical simulation results in graphical form via histograms or scatter plots (as illustrated in Fig. 3-15 and Fig. 3-16).

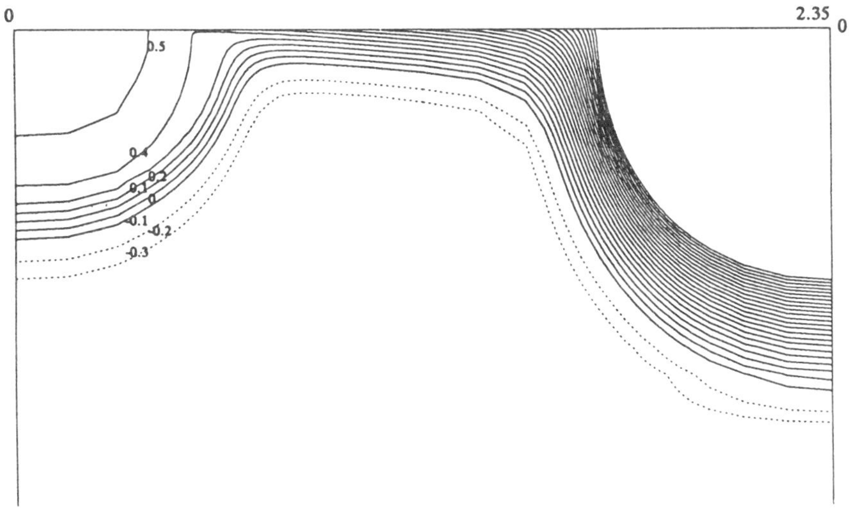

Figure 3-14: Equipotential contours.

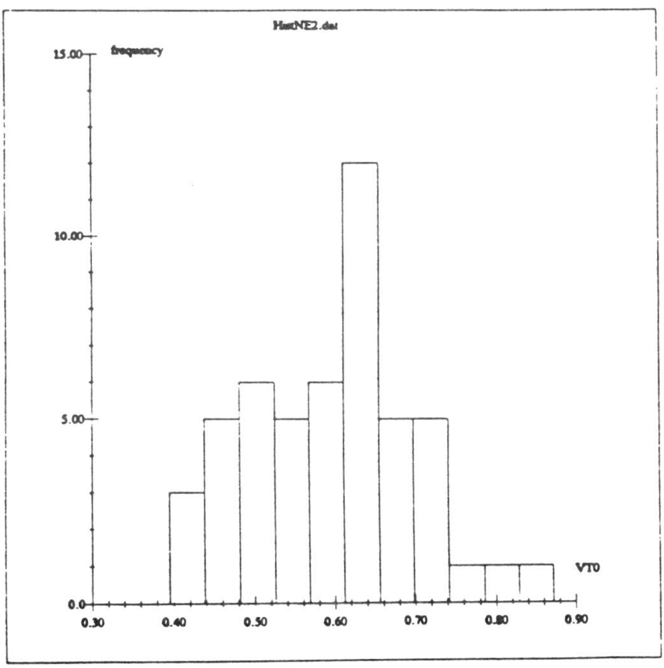

Figure 3-15: Histogram.

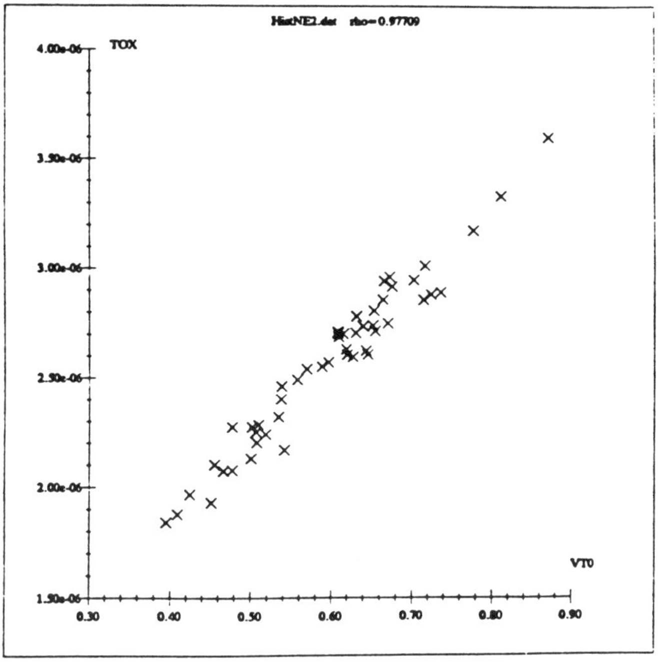

Figure 3-16: Scatter plot.

Tuning is accomplished by using PROMETHEUS with in-process and test structure measurements gathered from the fabrication line. PROMETHEUS generates the process disturbance files for a particular supervisor and is capable of performing *hierarchical tuning,* thereby accounting for the special correlations that occur within lots, wafers and chips. Although tuning is a computationally expensive task, FABRICS does not have to be re-tuned for different IC layouts or modified process parameters because the process disturbances in FABRICS are independent of process controls and device layout.

Once FABRICS has been tuned to a given fabrication line, it can be used to optimize the layout design rules such as minimum width, minimum spacing, or overlaps between different regions. The objective is to achieve a probability of failure for a given rule that is less than some threshold value. FABRICS can provide the statistical distribution of the edges of the geometrical objects of a layout by taking into account sources of variations as mask misalignment, etching fluctuations, lateral diffusion, depletion layer

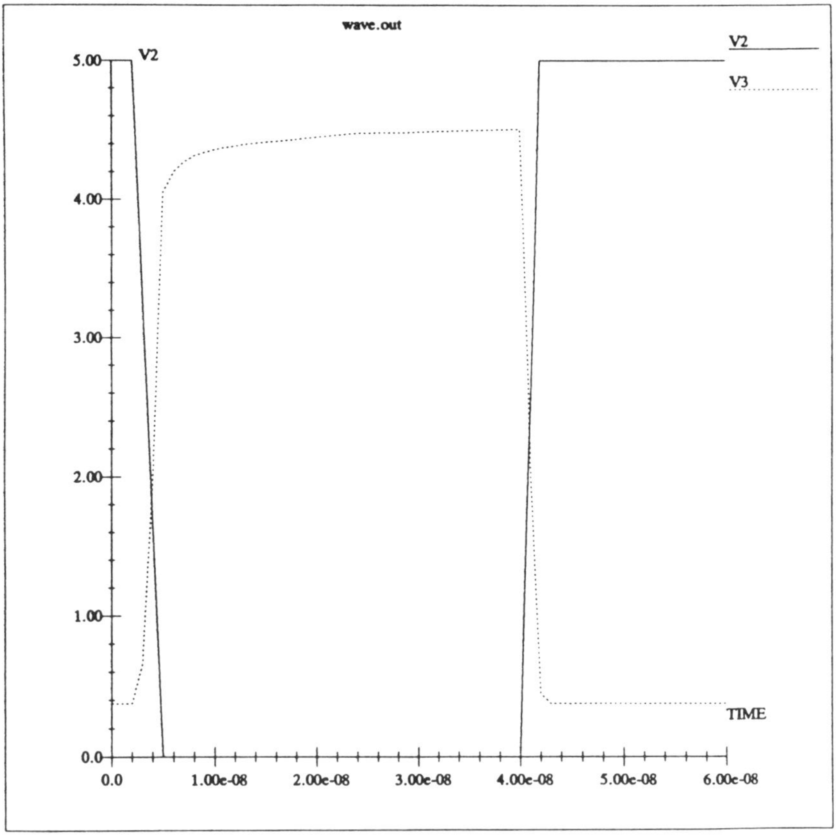

Figure 3-17: SPICE waveform.

expansion and the bird's beak effect. The statistical design rule developer STRUDEL [97] has been included in PEW to combine the information about the layout edge distributions from FABRICS with the information about spot defect distributions to optimize the design rules.

Finally, to observe the impact of the process design on circuit performance, PEW can access a circuit simulator such as SPICE and display simulation results as either waveforms or transfer curves. A waveform editor which allows for extraction of such circuit performances as delay times or power dissipation in the circuit is a part of the Workbench interface. Hence, the user can perform such activities as worst-case analysis or even statistical design centering in terms of either process control or device layout parameters.

3.4.5. Extensions

Current efforts focus on extending the PEW system to accommodate a variety of simulators. The problem of interfacing numerical device simulators has been solved by generating meshes in the regions of interest and specifying the impurity concentrations for all mesh points. This representation has been already used with the numerical device models for small-geometry MOSFETs [99]. At present, the direct interface to other simulators would still require a number of translators to generate the data files in a format acceptable to these tools. To alleviate this problem it is expected that a Profile Interchange Format (PIF), which is being developed by an EDIF subcommittee, will be used in our system.

Chapter 4
Statistical Analysis

This chapter describes several examples of the statistical analysis of IC performance. The first section is devoted to the description of the new approach to statistical timing analysis which offers a very significant improvement in the speed of evaluating the timing behavior of digital MOS IC's. This approach is especially well suited for the statistical design purposes such as design centering or worst-case design. The latter is the subject of the next section in which the worst-case analysis is formalized. A software system, WORCAN, is also described in this section together with some computational examples. Finally, the last section describes an approach to statistical optimization of the functional blocks of IC's. The optimization is carried out in terms of both layout and process control parameters. FABRICS is used in all the examples in this chapter to generate the data for statistical analysis.

4.1. Statistical Timing Simulation[1]

A key consideration in designing large CMOS chips is timing. Currently, the best means for verifying the behavior of VLSI circuits both from the logic viewpoint and from the timing viewpoint, is simulation.

Unfortunately, while strict hierarchical design methodologies have been developed in order to manage the complexity of the design process, the simulation algorithm has yet to take advantage of hierarchy. In many cases, only the simulation of the basic cells is performed, while the interaction of

---

[1]This section is based on the paper "A New Approach to Hierarchical and Statistical Timing Simulations" by J. Benkoski and A. J. Strojwas, IEEE Trans. on CAD of ICAS, vol. 6, pp. 1039-1052, 1987.

these cells and the interconnect timing are verified only by logic simulators and *timing verifiers*. In order to alleviate this problem, we need to find new ways of obtaining the timing data while verifying the logic behavior of the circuit. The association of larger chips with tighter timing constraints has increased the importance of both accuracy and efficiency. Furthermore, the need to verify large portions, if not the whole design, has definitely imposed a stress on hierarchical approaches. In addition, in today's technology, the assumption that process parameters are constant between dies, or even within a die, is no longer valid. As a result, statistical verification of the timing is now essential. Since statistical analyses require repeated runs, the cost of each simulation is even more crucial. Moreover, in order to perform a meaningful analysis of the process parameter variations, the simulation accuracy must be further improved. In this section, we will present a new methodology for timing simulation which was developed with the intention of satisfying both hierarchical and statistical requirements.

4.1.1. Overview

One obvious approach towards obtaining statistical timing data is to perform a Monte-Carlo experiment using a circuit simulator. However, such a technique requires a large number of simulation runs and often builds an unnecessarily complex model of the physical behavior of the circuit in order to compute the delay. A slightly better result in terms of computational cost can be obtained by employing an *importance sampling strategy* [103] that optimizes the selection of the simulation experiments. Unfortunately, even the reduced number of simulations that are necessary can become impractical. An alternative approach to timing verification involves associating a statistical delay model with each large block and using a *delay propagation operator* to compute the appropriate statistical output delay as a function of all of the input delays. In this way, a statistical model of the delay through a critical path could be easily computed. This idea was first implemented in TA [44] which used a statistical delay model based upon normal distributions characterized by their mean μ and variance σ. The delay propagation operator simply added the delay through the block to the latest of the input signals, the latest being defined by $\mathrm{Max}(\mu_i+\beta\sigma_i)$, where β is a *certainty factor*. Unfortunately, this algorithm introduces some anomalies. First, this algorithm produces only nominal distributions, and, as our results have shown (see Section 4.1.6) the actual distribution is not always unimodal. Furthermore, TA's algorithm chooses one of the distributions as the basis to compute the output distribution, and, of course, if the two distributions are very dissimilar this will result in a large difference between the computed output distribution and the actual convolution of the input distributions. A *best fit* delay propagation operator [128], in which the output delay is computed as the nominal distribution that fits the best the actual convolution of the input distributions was later proposed. However, while this solves the anomaly we have described above and approximates with good accuracy similar input distributions, it still does not allow us to model the non-unimodal distributions that would result from the convolution of significantly dissimilar input distributions. Moreover, the most important problem in the algorithm is the absence of correlation between the delays. There is no doubt that the variations in the delays are the results of common factors, and, particularly in the case of the process parameter variations, the correlation can be very high. By assuming statistically independent delay models, the probability of the worst case result is artificially increased, resulting in tight constraints that will eventually lead to overly conservative designs. Further work within this framework partially corrected the problem by introducing a correlation coefficient matrix [101] whose computation was eased in the limited library-based gate array environment. In this case, it is possible to compute the correlation between the delays of all of the gates in the library. This work also recognized the need to account for both intra-die and inter-die variations by extending the matrix to both on-chip and chip-to-chip correlation coefficients. Unfortunately, while this technique certainly opened new horizons, it is not general enough nor applicable to other environments.

4.1.2. Our Approach

As we have seen before, the absence of modeling of the logic behavior in timing verifiers has seriously limited their use by causing false paths to be discovered. In statistical problems, inevitably higher computational costs constitute another obstacle since the cost of the wasted effort associated with discovering false paths increases dramatically. In this section we describe an alternative approach that is based on a simulation paradigm. More specifically, as in circuit simulation, we propagate user-defined stimuli through the circuit and *simulate* its real response, but while circuit simulation tries to mimic as closely as possible the real, continuous (*analog*) waveforms, we will be concerned only with the more abstract *timing* models. Timing models simply characterize real waveforms by their associated logic transitions, the time of their occurrence, and some measure of their duration. This *transition* model can be formally defined in terms of the real waveforms:

- **starting time:** the beginning of a transition, its time of occurrence. It is measured as the time at which the real waveform crosses 10% of the voltage swing associated with the transition.

- **arrival time:** the end of a transition. It is measured as the time at which the real waveform crosses 90% of the voltage swing associated with the transition.

- **transition time:** the duration of a transition. It is defined as the difference between the *arrival time* and the *starting time*.

- **delay:** the delay between two transitions. It is defined as the difference between the *starting time* of the first transition and the *starting time of the second*. In the case of a delay through a block, it is simply the difference between the *starting time* of the output transition and the *starting time* of the input transition causing it.

A graphical interpretation of these definitions is shown in Fig. 4-1, in which the waveform has been approximated by a linear interpolation between the voltages corresponding to the starting time and the arrival time.

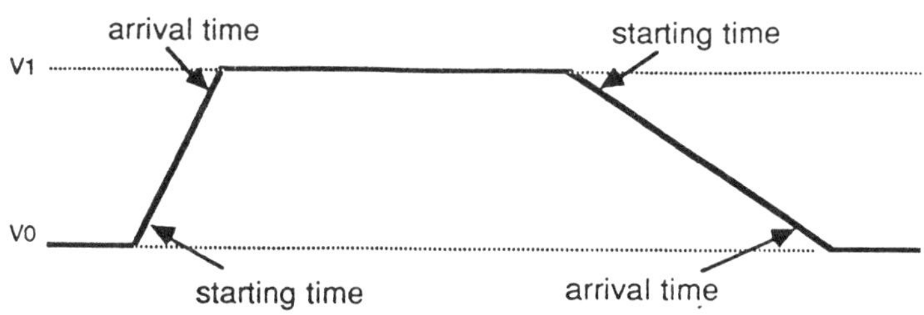

Figure 4-1: A waveform.

We wish to build a model based upon the *logic* and *timing* characteristics of the *output transitions* resulting from an *input transition*. Therefore, we need to partition our circuit in such a manner that the resulting blocks will have well-defined inputs and outputs. Most approaches to circuit partitioning involve flattening of the circuit, clustering of elements according to impedance, and identification of unique blocks. While these methods may be useful for circuit simulation, they are not consistent with the design hierarchy. Keeping in mind our goal of integrating the simulation process and the design process, we are motivated to perform partitioning according to the design hierarchy. Although the leaf cells of this design hierarchy usually contain very simple elements (single transistors, contacts, etc.), the higher levels contain basic logic gates and eventually *logically meaningful* blocks such as adder bit-slices, flip-flops, and control logic blocks. We will concentrate on those logically meaningful blocks, as they have well-defined inputs and outputs. In addition to the inputs and outputs, we will also require the *storage nodes* to be defined.

Our modeling methodology is based upon a dissociation of the *logic behavior* and the *timing behavior* of the blocks. The *logic behavior* of a block is described in terms of a set of *logic states*. A *logic state*, is defined in terms of voltages (or their equivalent logic values) at the input, output, and storage nodes, assuming the inputs are stable at one of the supply voltages and the block has reached its steady state. Being intended as a verification tool, STAT! models only the legal behavior of the blocks. Since usually not all of the possible combinations of input values are meaningful, some of them should therefore not appear in the normal operation of the block as it was intended by the designer. These illegal combinations are called *illegal states* and, although their appearance in the course of a simulation is detected and flagged, they are not part of our model. A simple example of an illegal state can be found in any block clocked by two non-overlapping clocks. the two clocks cannot be "ON" at the same time. An analogy can easily be made with the switching theory and will serve as the basis for the rest of our discussion. Each *logic state* can be viewed as a state in an equivalent finite state machine (FSM). The resulting state graph includes only the *legal states*. For each of the legal states we define the set of input changes (transitions) that can appear in the normal operation of the block[3]. These input transitions cause associated output and storage node changes which, once the block has reached its new steady state define another *legal state*. In the state graph, these transitions from state to state are represented by their equivalent edges. Fig. 4-2 shows the FSM representation of a half-adder.

By observing the finite state machine representation that we have just described, it can be seen that, within this model, the timing behavior of the block can be modeled by the response of the output and storage nodes to a

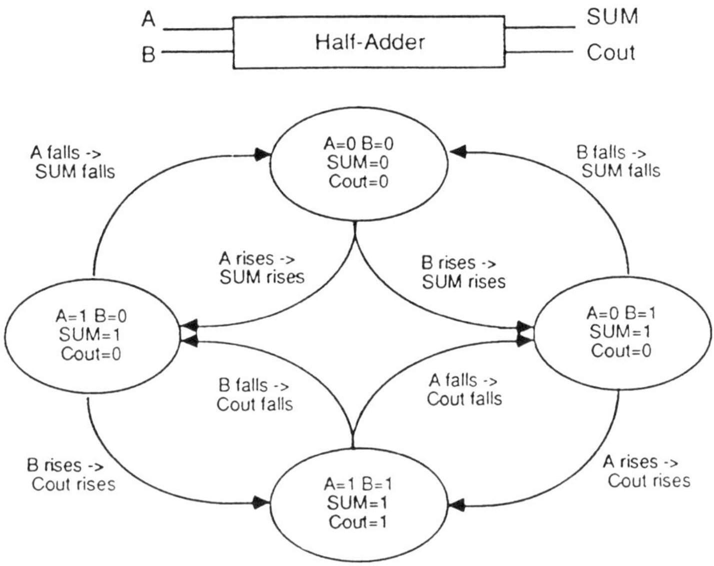

Figure 4-2: FSM representation of a half-adder.

change in the inputs. In terms of our timing level definitions, this amounts to finding the output and storage node transitions resulting from an input transition. Since a transition is fully characterized by its *starting time* and its *transition time*, we define a block timing model that consists of two components:

- **Delay:** the delay through a block. The output transition starting time can be computed as the sum of the input transition starting time and the delay.

- **Output transition time:** the transition time of the output transition. The output transition arrival time can be computed as the sum of the output transition starting time and the output transition time.

The experience of timing verifiers with models that did not account for the input slope effects on the delay, as well as the results of our own experiments, have influenced us to build those models as a function of the input transition time. Furthermore, the resulting model is self-consistent as it takes a transition defined by its starting time and transition time and outputs another transition defined in the same terms. In order to make our model applicable to many different situations, we have also incorporated the output capacitance as a parameter. Thus, a single block timing model can be used in different instantiations of the corresponding cell without introducing any error as a result of the different load it is facing. Moreover, this fits very well in our philosophy of integrating the simulation process and the design process, as neither the cell nor the model need to be changed to accommodate different environments. A complete description of the method for building this timing model as well as a more refined model for large blocks can be found in Section 4.1.3.

Our goal in this work was to easily evaluate the effect of the process parameter distribution upon the timing behavior of a circuit. In addition, we wanted to find a way to account for the correct correlation between the delays in order to avoid the conservative results obtained in the past. Both goals were achieved by extending the nominal model and building it as a function of statistically independent process parameters. It is important to stress that, when using this statistical model, a different *delay* and a different *output transition time* are obtained for different values of the process parameters. Since the model is based upon statistically independent process parameters, the variations in the computed delay and output transition time for different blocks are automatically correlated. Therefore, when running a simulation for certain values of the disturbances in the process parameters, the computed timing is based upon delays that implicitly contain the correct correlations.

Although it is important that the parameters be independent, their actual identity and number are not part of our methodology. In the current implementation, we model variations in the polysilicon line width which defines the transistor length L, the active region width which defines the transistor width W, and the metal line width W_m which plays an important role in variations of the interconnect capacitance. This set of parameters represents a good approximation of the first-order effects of process disturbances. As an example of the need for low-level, statistically independent parameters, let us consider the problem of modeling of the variations in the threshold voltage V_T. Because V_T is strongly related to the transistor length in short channel transistors, it is not statistically independent of L. Therefore, we need to find lower-level parameters which influence V_T and are independent of all the other parameters. In this particular case, we could choose to add N_s, the surface concentration. By doing so, we enforce an interesting decomposition of the problem, and we separately model the independent effects. The transition-based model that was defined above lends itself well to an event-driven propagation scheme. Indeed, while the block *logic model* allows us to determine the output and storage node transitions resulting from an input transition, the block *timing model* enables us to compute the starting time and output transition time of those resulting transitions on the basis of the input transition characteristics. In other words, the association of both models provides us with an operator that fully defines the output and storage node transitions caused by an incoming input transition. By defining an event as a transition, we can construct a framework in which the simulation algorithm is made very simple:

1. Take the first event (transition) in the event queue.

2. Find resulting output and storage node transitions in the block logic model.

3. Using the incoming transition starting time and transition time, compute the starting time and transition time of the resulting transitions from the block timing model.

4. Schedule the resulting transitions in the event queue.

5. If the event queue is empty, terminate. Otherwise go back to 1.

In the *nominal mode*, the block models are called with the nominal values of the process parameters, and the result of the simulation gives us the *nominal timing*. Of course, if we have built the statistical model , we can call the block models with disturbed values of the process parameters and compute the resulting, correctly correlated, *disturbed timing*.

4.1.2.1. Timing reevaluation

The STAT! approach, as described until now, seems to be an appropriate way of quickly computing the *nominal* as well as the *disturbed* timing data. However, when facing massive computation requirements (as in the computation of the probability distribution function of a delay), the algorithm described so far will eventually succumb. The proposed methodology goes further and includes a *statistical mode* in which we can reevaluate much faster the timing characteristics of the circuit for disturbed values of the process parameters. This *statistical mode* is based upon a single *nominal* simulation followed by fast reevaluation of the resulting timing. In the flow chart shown in Fig. 4-3, this first *nominal* simulation is represented by the black arrows. As a result of the first nominal timing simulation, we have built an event queue of all of the events that have happened in the circuit during the simulation. This queue represents a model of the ordering of the events according to both *chronology* and *causality*. Indeed, not only is the queue ordered according to increasing *starting time*, it also includes the *causal* relationship between the output and storage node transitions and the input transition that caused them (see Section 4.1.6). At the end of the nominal timing simulation, the timing data of the circuit can be reevaluated by recursively recomputing the starting time and the transition time of all the events according to the *causality tree*. In Fig. 4-3, this reevaluation is represented by the grey arrows.

This fast reevaluation mechanism is based on the fact that we know that each output event results from some input transition. Therefore, there is a timing model that has allowed us to compute the *delay* and the *output transition time* of the event. Under the new, disturbed values of the process parameters, the same timing model will return other values, therefore causing new *starting time* and *transition time* to be computed for the resulting transition. Of course, unless this transition was user-defined and applied to a primary input, it was also modified by the disturbed values of the process parameters, and, therefore, we need to *recursively* recompute the timing of all transitions. At the end of the complete reevaluation, we verify that the ordering of the events at each block has not been changed. If this is the case, we can be sure that the recomputed timing behavior is correct. However, if we find that the ordering is modified, this may indicate that, under these particular values of the process parameters, a *parametric fault* has occurred.

4.1.3. Characterization

By *characterization* we refer here to the process in which we observe the behavior of the block under different conditions, *characterize* this behavior,

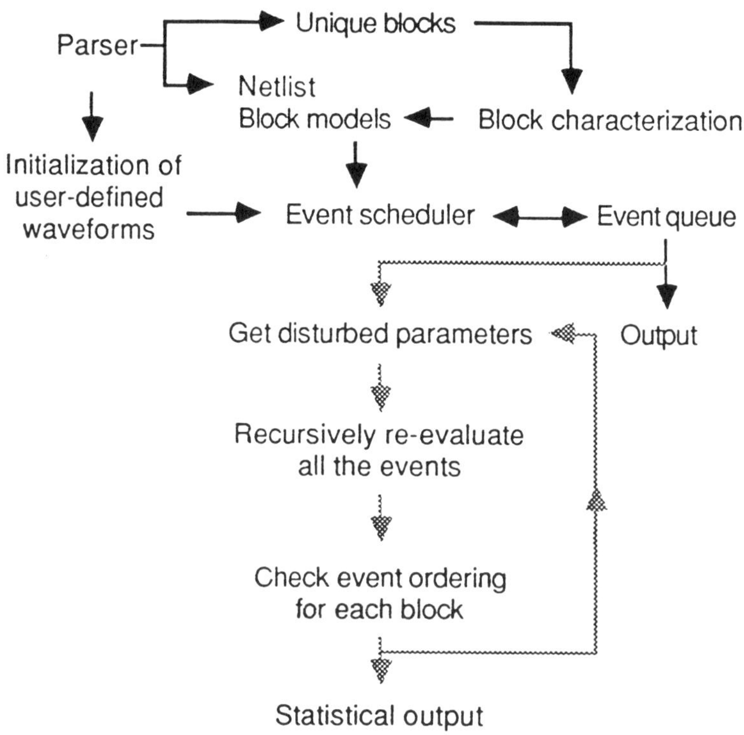

Figure 4-3: The flow chart.

and build a simplified model that will exhibit similar characteristics. In the past, many models of the timing behavior have been proposed. Most have tried to generate a simpler model of MOS logic gates in terms of equivalent resistances, capacitances, and controlled current and voltage sources [93, 75]. While these approaches provide a significant speed-up when compared to circuit simulation, they still require an analysis of the resulting circuit. Other works departed from this approach by restricting the type of input and output waveforms to a set of predefined functions and providing a model to compute the output waveform as a function of the input waveform [84]. This macro-model does not require any additional analysis and lends itself very well to the event-driven simulation schemes. Our model is based on the same principle, except that we consider a much simpler approximation to the waveforms, requiring only the modeling of the *delay* and the *output transition time*. As we mentioned in Section 4.1.2, the *nominal model* is a function of only the *input transition time* and the *output capacitance*, while the *statistical model* is also a function of the process parameter variations.

In order to build this parametrized timing model, we have implemented the following scheme:

1. Perform several circuit simulations with different values of the parameters included in the model.

2. Extract the relevant data from each simulation (i.e., delays and slopes).

3. Build the model by linear regression.

In the current implementation, circuit simulation is performed for each combination of the three values of each parameter: the two predicted extreme values and the mean value. While the resulting full factorial plan is straightforward, it is certainly not the most efficient way of building a model by experimentation and was chosen only in order to allow an easy testing of our methodology. Various methods of reducing the number of experiments required in order to build parametric models (such as fractional factorial experiment plans [10]) are employed.

During the course of this research, we have tested both SPICE2 [80] and an early version of CINNAMON [123]. We have found that the accuracy provided by CINNAMON was sufficient and the resulting speed-up very significant. Typical CPU-times for a full characterization were of the order of 5 minutes for CINNAMON and about 1 hour for SPICE2 on a VAXstationII.

Using the timing data computed from the circuit simulations, a parametric model is built by linear regression [10]. Here again, the methodology does not specify the degree of the polynomial used in the model. We have found however that a second order polynomial allows to obtain a high level of confidence in the model computed by the linear regression program. Furthermore, many experiments have shown that the *delay* and the *output transition time* are almost proportional to the *input transition time* (see Fig. 4-4) and to the *output capacitance*, and the non-linearities can easily be accounted for in a second order polynomial model.

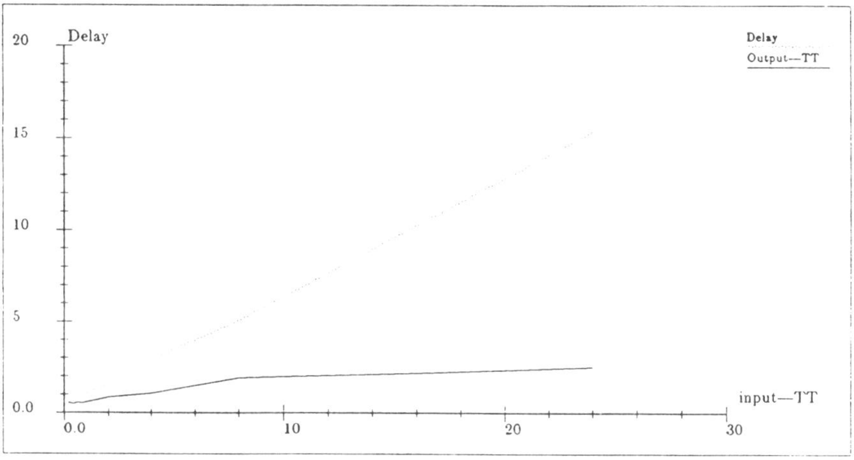

Figure 4-4: Influence of the input transition time.

4.1.4. Delay decomposition

The delay model we have described so far can be applied to any type of block without any restriction as to its size or functionality. However, this model can be further refined when considering very large blocks or blocks with special topologies. The idea behind this refinement is to isolate the influence of the *input transition time* and/or the *output load*. If we consider, for instance, a very long inverter chain, it can be shown experimentally that if the range of the *input transition time* is bounded, there is a number N such that the *transition time* of the Nth inverter is independent of the *input transition time*. Therefore, the *delay* between the $(N+1)$th inverter and the output will be independent of the *input transition time*. Similarly, in CMOS, the influence of the *output load* is usually limited to the last gate, and, therefore, the *delay* up to that gate is independent of the load. Whenever such an isolation is possible, we decompose the *delay* from the input to the output into three delay sub-components:

- **the input delay** - τ_{inp}: computed as the delay through the part of the block which was found to be sensitive to the *input transition time.*

- **the output delay** - τ_{out}: computed as the delay through the part of the block which was found to be sensitive to the *output load.*

- **the intrinsic delay** - τ_{int}: computed as the delay through the part of the block which was found to be insensitive to the *input transition time* and to the *output load.*

The delay through the block is simply computed as the sum of the three delay sub-components, and we can write the following equation:

$$\tau_{io}(I_{tt},L) = \tau_{inp}(I_{tt}) + \tau_{int}() + \tau_{out}(L),$$

where τ_{io} is the delay from the input to the output, I_{tt} is the input transition time, and L is the output load.
Of course, if such a valid decomposition exists, we can also write:

$$O_{tt}(I_{tt},L) = O_{tt}(L),$$

where O_{tt} is the output transition time function.
The determination of the *output delay* can be realized by a simple experiment which can be included in the characterization plan as it is composed of these simple steps:

1. Simulate the block for the smallest capacitance in the considered range

2. Simulate the block for the largest capacitance in the considered range

3. Compare the *transition times* at each of the nodes

4. Include in the *output delay* all the components whose *transition time* differ by more than a given threshold value.

The problem of the *input transition time* isolation is somewhat more complicated as, in most technologies and for most gates, there is an influence of the input transition time on the output transition time. However, there are still many cases in which an *input delay* can be determined. The first class of blocks includes all the long chains of combinational logic in which the high gain of the gates restricts the influence of the *input transition time* to the first part of the block. In this case, a experiment similar to the one described above can be performed. We have found, for instance, that in a ring oscillator fabricated in a MOSIS 3-micron CMOS process, the sensitivity to the *input transition time* was limited to the first two gates, for *transition times* between 0.5 and 25 nanoseconds.

The second class of decomposable blocks deals with blocks in which, in the normal operating conditions, the logic specifications limits the influence of the *input transition time*. The semi-dynamic master-slave flip-flop shown in Fig. 4-5 is a typical example of this class. In this case, the logic specifications do not allow Phi1 and Phi2 to be "ON" at the same time and, therefore the sensitivity to the *input transition time* is limited to the first transmission gate and inverter.

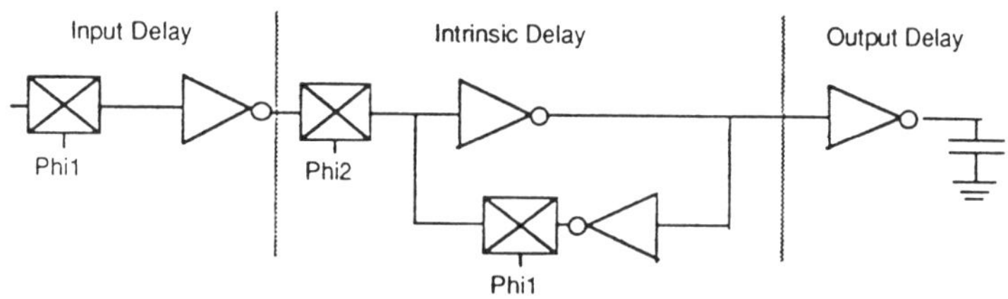

Figure 4-5: A semi-dynamic master-slave flip-flop.

While during the first stage of the simulation process, we do want to see all of the nodes in order to verify the behavior of the circuit, once the designer is assured that the blocks function properly, the intermediate nodes become of less interest and the idea of *concatenating* the blocks becomes appealing. The delay decomposition has allowed us to develop such a *block concatenation* scheme. Let us assume that we have already fully characterized two blocks and decomposed their *delays* into the three sub-components. Since we know the configuration of the circuit, we also know the *output load* of the first block. This load is composed of the sum of the capacitance of the output node and the input capacitances of all the blocks connected to it. Using this information, we can now replace the load parameter in the τ_{out} and the O_{tt} functions of this first block and compute the values of the *output delay* and *transition time*. This *transition time* is, of course, the *input transition time* of the second block and replaces the corresponding parameter in its τ_{inp} function to allow us to compute the input delay. This transfer of information between the two blocks is shown in Fig. 4-6 which also demonstrates the last part of the algorithm: the two blocks can now be concatenated and the *intrinsic delay* of the resulting block is composed of the sum of the *intrinsic delays* of the two blocks, the *output delay* of the first block, and the *input delay* of the second block.

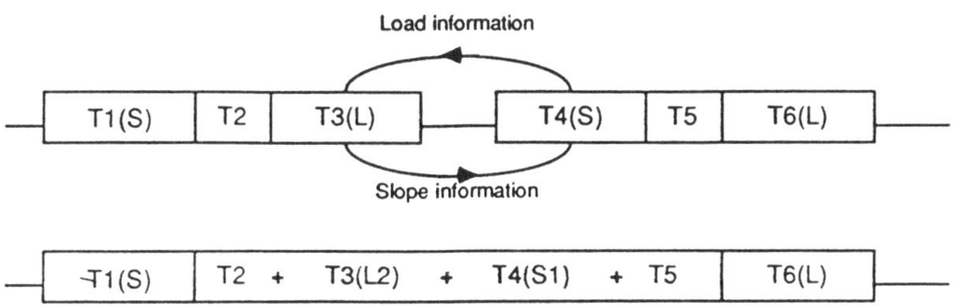

Figure 4-6: The concatenation algorithm.

4.1.5. Nominal Simulation

The simulation relies upon an event-driven algorithm. We define an event as either the beginning or the end of a transition. The beginning event is called *start* event, and is scheduled at the starting time while the end event - henceforth referred to as *end* event - is scheduled at the arrival time. As a signal (i.e., transition) reaches a block, we must check whether it is going to cause the scheduling of a transition at one of the outputs and/or a storage node. The logic model will provide this information while, if there is a resulting transition, the timing model will allow us to compute its starting time and arrival time. Fig. 4-7 shows the data flow in the event scheduler.

Unfortunately, the situation described above is only an ideal picture. The following cases may occur and cannot be dealt with by the simplistic algorithm described above:

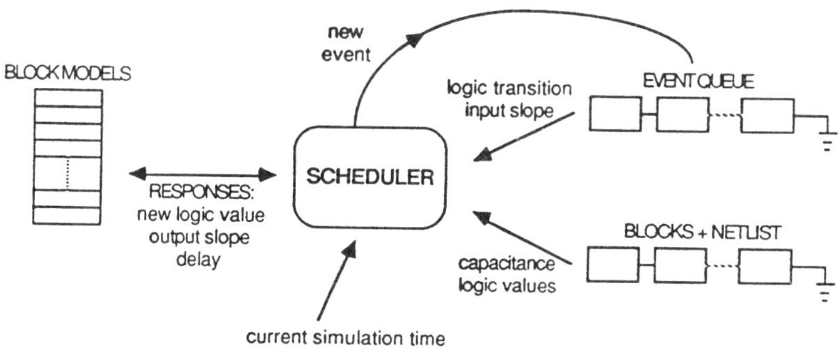

Figure 4-7: Data flow in the event scheduler.

1. Conflicting inputs may cause a glitch (i.e., an incomplete transition) at the output.

2. Fast inputs may cause the block to temporarily enter an illegal state.

3. If a block input changes while another one one is still in the midst of a transition, a resulting output response might be corrupted.

As a result, we cannot assume anything about the outcome of a transition until it completes. It is not until a transition completes, i.e., the simulator internal time reaches its arrival time, that the new value is entered in the block logic state. At this point, the legality of the new state is checked. If it is found to be illegal, the simulation backtracks to the previous legal state and, using our knowledge of the completion of the transition, re-process the event queue. In pathological cases, there can be multiple iterations until convergence is reached. Fig. 4-8 shows the iteration process which occurs as a result of a combination of cases 2 and 3. Each box represents an event in the queue. The values at the top of the box are the block logic values, and the arrows show the relationship between the scheduled event and its cause (note that the *start* event schedules its corresponding *end* event). In this picture, A rises and causes no change at the output. Then, B starts rising

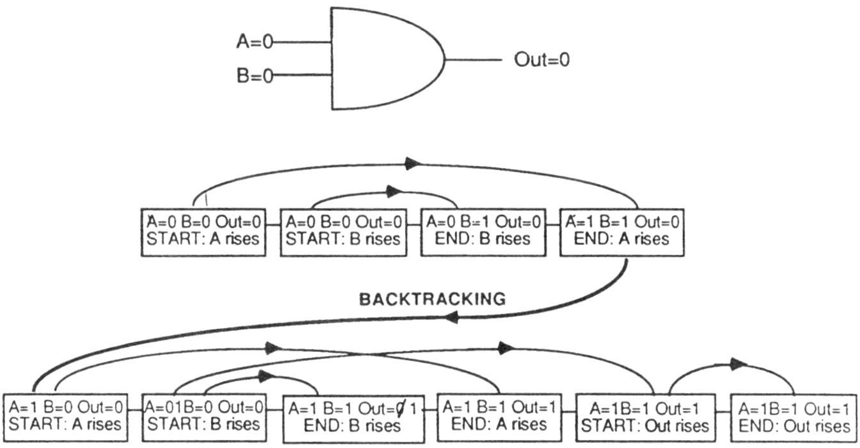

Figure 4-8: An iteration scheme.

but since A has not completed yet, B does not induce any output change either. When B completes (third box), its new value is entered in the block logic state. However, as A completes and its value is entered, we find that the block is in an illegal state. The simulation then backtracks to the first box, and the knowledge of the completion of A allows B scheduling of the correct output response. In this special case, we notice that the illegal situation is appearing again in the third box, however, this is recognized as being temporary (i.e., the output change is already in the queue), the situation is corrected, and the simulation proceeds.

The simulation algorithm has been implemented in a software tool called STAT!. Below we present some simulation examples. Fig. 4-9 shows a simple test circuit and its decomposition into blocks. In this case, the models were obtained from CINNAMON. Table 4-1 shows the comparison between the delays computed by our simulator (column I), a full simulation with CINNAMON (column II), and a full simulation with SPICE2 (column

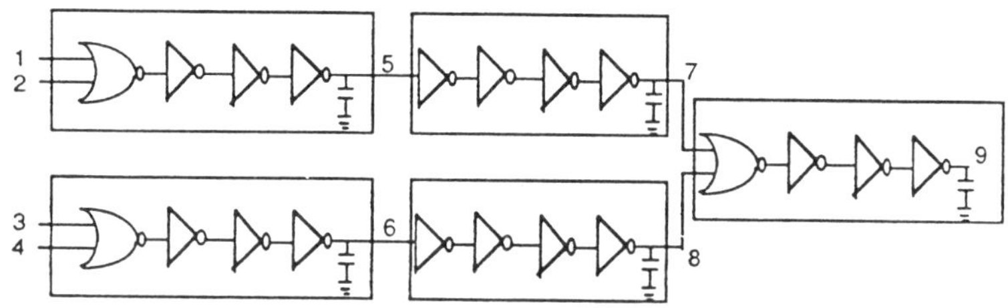

Figure 4-9: A test circuit.

DELAY	I	II	III	Δ%(I-II)	Δ%(I-III)
Node 4 -> Node 6	4.89 ns	4.63 ns	4.41 ns	+ 5.6	+ 10.9
Node 4 -> Node 8	11.5 ns	10.9 ns	10.6 ns	+ 5.5	+ 8.5
Node 4 -> Node 9	17.2 ns	16.2 ns	15.3 ns	+ 6.1	+ 12.4

Table 4-1: Comparison of STAT! (I), CINNAMON (II) and SPICE2 (III).

III). The CPU requirements were 1.7 seconds, 53.0 seconds, and 439.0 seconds, respectively. The comparison with SPICE2 shows relatively high errors, but the comparison with CINNAMON sheds more light upon their origin: only about half of the error is to be attributed to our simulation method, while the other half is a consequence of circuit simulation errors during the characterization. The timing error incurred by our simulator is conservative and can be explained by the simple modeling of the input capacitance which, in the current implementation, is approximated by the constant gate capacitance.

DELAY	SPICE2	STAT!	Δ%
Carry-in -> Carry-out1	2.09 ns	2.27 ns	+ 8.7
Carry-in -> Carry-out2	4.88 ns	5.13 ns	+ 5.2
Carry-in -> Carry-out3	7.81 ns	8.04 ns	+ 2.9
Carry-in -> Carry-out3	10.6 ns	10.9 ns	+ 2.6

Table 4-2: Delays for the 4-bit adder.

The waveforms produced by STAT! (using a 1-bit slice model obtained from SPICE2) and by SPICE2 for the carries of a 4-bit adder are shown in Fig. 4-10. STAT! required 1.2 CPU-seconds, while SPICE2 needed 392.7 CPU-seconds. Table 4-2 shows the comparison of the delays computed by our simulator and by SPICE2. Again, it can be seen that our results are slightly conservative. In this case, the model was characterized by SPICE2, suppressing the effect of the inaccuracy of the characterizing simulator and significantly reducing the errors.

Fig. 4-11 shows how glitches are handled by STAT! and it can be observed that the peak voltages as well as the timing of the glitches are very similar to those produced by SPICE2. The CPU requirements were 1.9 CPU-seconds for STAT! and 391.5 CPU-seconds for SPICE2.

4.1.6. Statistical Simulation

The *statistical mode* of STAT! is based upon a single *nominal* simulation followed by many, much faster, reevaluations of the circuit timing for disturbed values of the process parameters. As opposed to the cases in which the nominal simulator is used as a stand-alone tool and relies on the nominal timing model, in this case we will perform the nominal simulation on the basis of the *statistical model*, evaluated for the *nominal values* of the process parameters. Fig. 4-12 shows a typical example of an *event queue* built by this first *nominal* simulation. The resulting data structures include not only the *event queue* itself, but also the *causality tree* which relates each event to the event that caused it, and, for each *start event*, a pointer to the *timing model* that was used to compute its *starting time* and *transition time*.

Thus this *event queue* contains the ordering of the events that happened in the circuit during the simulation, as well as the cause to each of them and the models that served in order to compute the timing data. Since those models are built as a function of the process parameters, if we want to alter the values of these process parameters, it is possible to recompute all of the timing data by simply reevaluating all of the models attached to the *start events*. Since the *delay* and the *transition time* computed by each of these models are functions of the *transition time* of the transition that caused their associated responses, that *transition time* is required in order to recompute the timing data. Fortunately, by climbing the *causality tree* from the roots to the leaves, the timing data of the *causing* event will always be

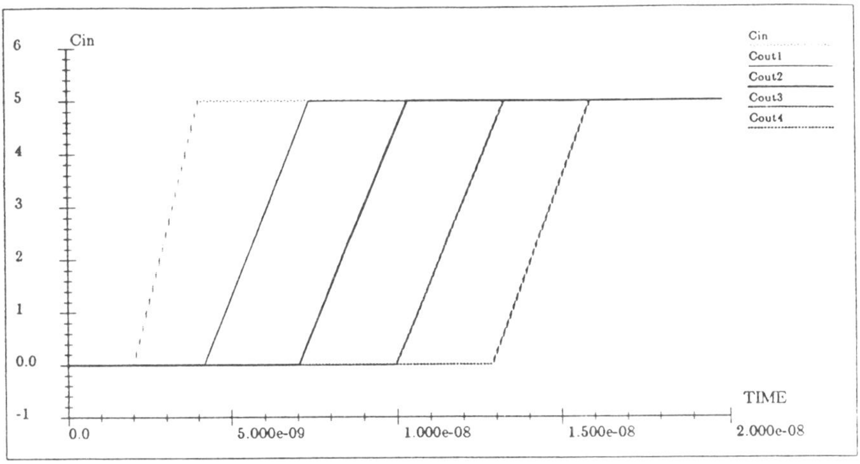

(a) STAT! simulation

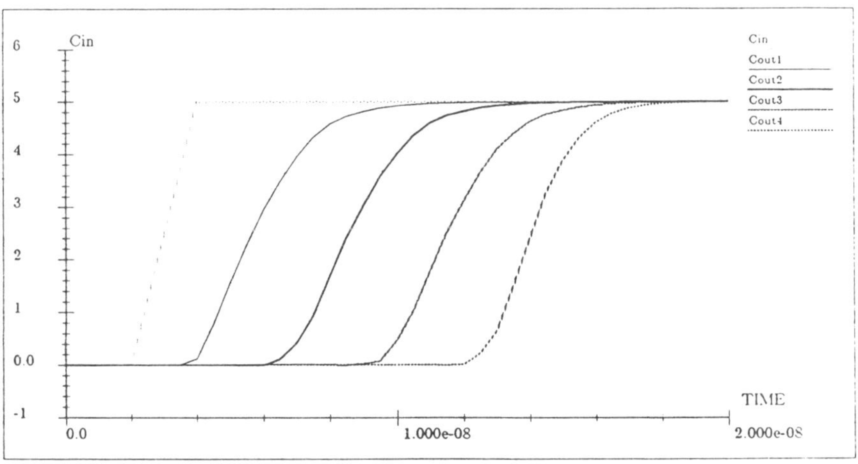

(b) SPICE2 simulation

Figure 4-10: Waveforms for the 4-bit adder.

computed before that of the *resulting* event. This idea is actually implemented in the *statistical mode* of STAT! by recursively reevaluating the *starting time* and the *transition time* of the events according to the *causality tree*, starting from the leaves.

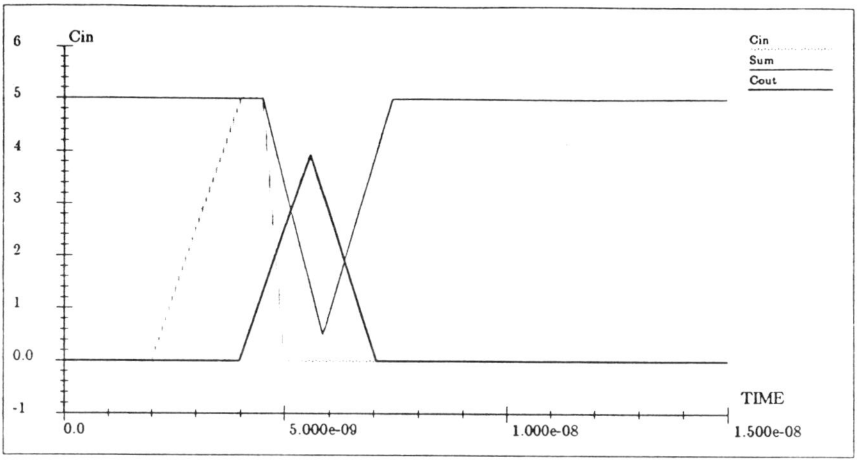

(a) - STAT! simulation

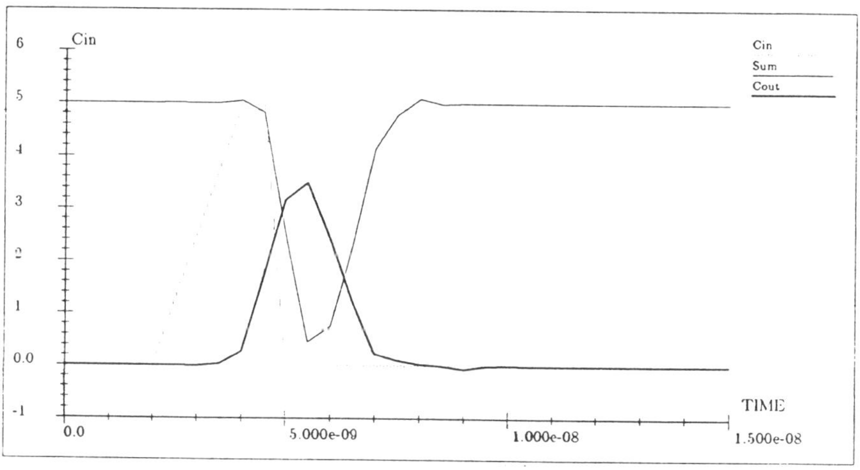

(b) - SPICE2 simulation

Figure 4-11: Glitch in the adder.

As a result of this reevaluation, the *starting time* and the *arrival time* of each transition are modified, possibly causing a modification of the ordering of the events in the *event queue*. While this modification is not important as long as it concerns unrelated events, it implies a different sequence of input

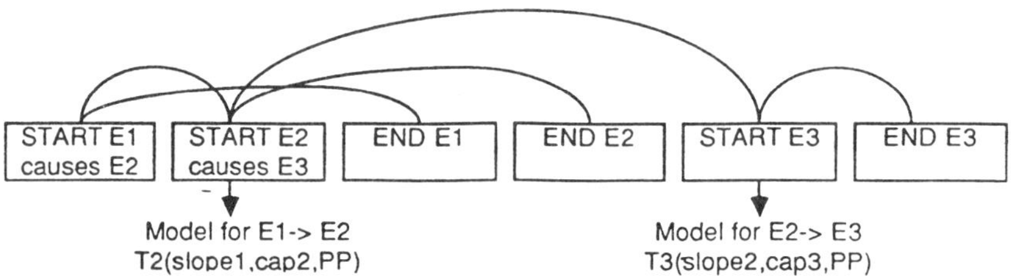

Figure 4-12: The event queue built by the *nominal* simulation.

signals and thus different output responses, when the ordering of the signals related to a same block is changed. Therefore, at the end of the recomputation of the timing data, the ordering of the events is checked on a block by block basis. If the ordering of the event is found unchanged, the timing data that resulted from the reevaluation is valid. Otherwise, if the new ordering causes a different output response, it simply means that under the given values of the process parameters a *parametric fault* has occurred in the circuit. If the output response remains the same, although it is important to advise the designer that the input sequence is changed, this fact should be detected and the simulation continued. Unfortunately, in the current implementation, STAT! is not able to detect this last case, and the simulation is aborted.

The important features of STAT! from the statistical viewpoint are thus:

- Implicit correlation of the computed delays
- Fast reevaluation of the timing data for disturbed values of the process parameters
- Detection of *parametric faults*.

Consequently, the general area of application will include all of the problems that require repeated computations with slightly different values of the process parameters in order to predict statistical data.

Computing the probability distribution function (PDF) of a delay is one of the most important tasks in the field of statistical timing analysis. For instance, from the PDF of the delays in the critical path, we can deduce the dispersion of the maximum frequency in the manufactured circuits. By computing several PDFs for different values of the variance of the process parameter distributions, we can predict the influence of tighter control procedures on this maximum frequency. Of course, in this problem, the correct correlation of the delays is crucial in order to perform a meaningful analysis. The usual approach in the past has been to build a Monte-Carlo experiment based upon a circuit simulator. As we already mentioned in the introduction, this method, although it can be slightly improved by adopting the *importance sampling strategy,* should really be the last resort as it is the most expensive and completely neglects the information accumulated by previous simulations. Of course, replacing the circuit simulator by the *nominal* simulator built into STAT! would already provide a significant speed-up but the approach that can be taken in the *statistical mode* is much more efficient. In this case, as we explained in the first section of this chapter, we will run a single *nominal* simulation at the beginning and each subsequent sample will simply require the reevaluation of the *event queue.* Although the *importance sampling strategy* can be adopted in this case as well, it has not yet been implemented. Fig. 4-13 shows the PDF of the delay through the test circuit shown in Fig. 4-9 computed by a Monte-Carlo experiment of 500 samples.

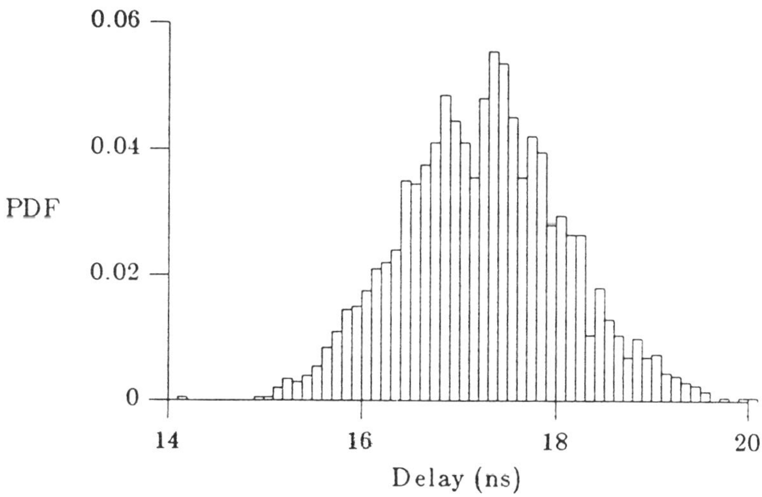

Figure 4-13: Delay PDF through our test-circuit.

The varying parameters were the transistor width W, the transistor length

L, and the metal line width W_m. The mean and the variance were respectively 3 microns and 0.2 microns, 3 microns and 0.2 microns, and 6 microns and 0.7 microns. The *statistical mode* of STAT! required a total of 17.5 CPU-seconds which is composed of roughly 10 CPU-seconds for the first *nominal* simulation and 0.015 CPU-seconds for each of the *event queue* reevaluation. The same experiment, if based on the *nominal* simulator only would have thus required 500 times the *nominal* simulation CPU-requirements, 5000 CPU-seconds. The same experiment, if based on SPICE, would have required 500 x 3000 CPU-seconds = 1,500,000 CPU-seconds, which is roughly equivalent to 17 days. The resulting PDF in itself offers an interesting indication as well: as it is not unimodal, it could not have been obtained by any of the *statistical timing verifiers* described in the introduction.

4.2. An Improved Worst-Case Analysis Procedure[2]

While it is desirable to perform a parametric yield prediction for a design prior to manufacture, in reality most circuit designers verify their design under some *worst-case* conditions. If IC performance under these conditions is still acceptable then the design moves into production. This procedure, called *worst-case analysis* , has been traditionally performed in terms of the electrical parameters of IC devices, such as threshold voltages and transconductances for MOSFET's. Such parameter sets are typically derived from exhaustive characterization of test structures, and are costly to produce. Moreover, these parameters are statistically dependent (correlated) random variables with a multilevel structure of variance (intra-die and inter-die) . In traditional approaches to worst-case analysis, the correlation coefficients between device parameters are not taken into account, and therefore IC performances are estimated for some *unrealistic* combinations of the device parameters.Hence the results of such an analysis are usually too pessimistic [25]. In this section we describe an improved approach to worst-case analysis that yields more realistic estimates of variations in device and circuit performances. This approach is based upon employing FABRICS II coupled to a circuit simulator. We also describe a software

---

[2] This section is based on the paper "A Methodology for Worst-Case Analysis of Integrated Circuits" by S. R. Nassif, A. J. Strojwas and S. W. Director, IEEE Trans. on CAD of ICAS, vol. 5, pp. 104-113, 1986

package that can be used to automate worst-case analysis and present some examples.

4.2.1. Worst-Case Analysis Methodology

Typically, the behavior of an IC is defined in terms of a number of performances (e.g. average power dissipated in the IC or inertial delay of a signal). Since each performance exhibits different sensitivity to process variations, a strict worst-case analysis should be carried out separately for each IC performance. However, because many of the performances are correlated, worst-case analysis can in fact be performed for a small subset of performances. Furthermore, since large digital IC's are composed of a relatively small number of different *cells* which are repeated many times within a chip, performances of these basic cells are very accurate indicators that can be generalized to the chip performance levels. Even the performances which are affected by the intra-die variations can be analyzed for small functional blocks (e.g. sense amplifiers in RAM's). If the timing of an IC design is of main concern, a critical path of a circuit can be considered to include the dominant parasitic elements associated with the interconnections. Hence, for most practical circuits we can restrict our attention to subcircuits of such a complexity that a worst-case analysis can be performed quite efficiently.

The choice of a subset of the independent random disturbances for worst-case analysis can be made based on the *sensitivity* of circuit performances to each disturbance. Thus, given a set of performances P_i, $i=1,..,n_p$, and a set of disturbances D_j, $j=1,..,n_D$, the normalized sensitivity of P_i to disturbance D_j can be written as:

$$s_{ij} = \frac{\delta P_i}{\delta D_j} \frac{D_j}{P_i} \tag{4.1}$$

We would like to use the sensitivity s_{ij} in order to (1) determine whether we need to consider the disturbance D_j in calculating the worst-case for performance P_i; and (2) to find the *direction* in which D_j must be adjusted to cause the performance to become *worse* with respect to *constraints* placed on that performance. For example, if we have a constraint on the delay through a combinational logic block of the form $t_{delay} \leq t_{max}$ then *increasing* t_{delay} would be equivalent to making it worse. We will refer to these directions as ***worst-case directions*** , such that adjusting the nominal value of a disturbance D_j in the worst-case direction (i.e. increasing or decreasing D_j) would cause P_i to become worse.

A local estimate of s_{ij} can be calculated by perturbing D_j by ΔD_j, and calculating the change in performance ΔP_i, thus:

$$s_{ij} \simeq \frac{\Delta P_i}{\Delta D_j} \frac{D_j}{P_i}. \tag{4.2}$$

Since Eq. (4.2) gives only a *local* measure of sensitivity, it cannot be used to determine a worst-case direction in cases where the relationship between P_i and D_j is non-monotonic. Furthermore, even in case where the relationship is monotonic, but highly non-linear, local measures of sensitivity may be misleading when trying to determine whether we need to consider disturbance D_j in calculating the worst-case of performance P_i. Fig. 4-14 shows two examples in which local measures of sensitivity can result in significant errors.

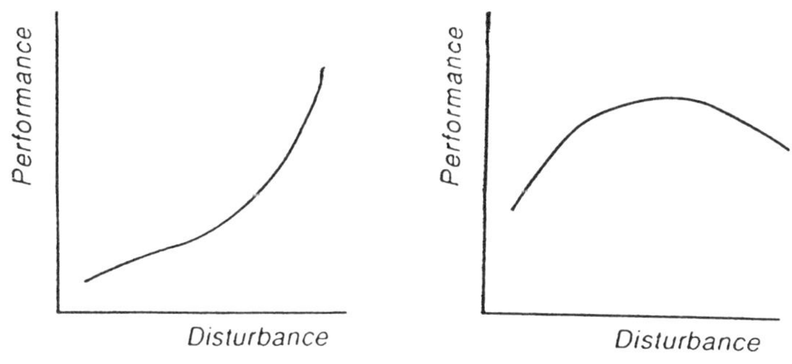

Figure 4-14: Performance dependence on disturbances.

In order to get a better estimate of sensitivity, we examine the behavior of P_i over a wide range of values of D_j. In particular, we calculate P_i for a number of values of D_j, ranging from D_j^{min} to D_j^{max}. At this point we have a table of values for the disturbance D_j and corresponding performance P_i. We can use such a table to determine whether or not a disturbance needs to be included in the calculation of the worst-case for a performance P_i based on the relative effect that D_j has on P_i. We can also use such a table to determine the worst-case direction if the relationship is monotonic, as our experience has shown most to be. For cases where the relationship is non-monotonic, we generate regression polynominals fitting P_i to D_j over the interval D_j^{min} to D_j^{max}, and use such polynomials to find *regions* where the relationship is monotonic, and thus identify one or more worst-case situations. Fig. 4-15 shows an example of how such a polynomial may be used.

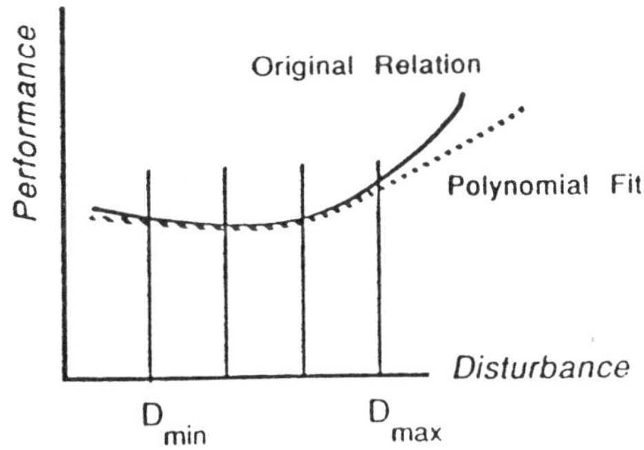

Figure 4-15: Using polynomial models.

Once the sensitivities of all performances to all disturbances have been calculated, and the worst-case directions for all the disturbances are determined, the worst-case for the performance may be calculated by adjusting *all* the disturbances in their worst-case directions. We are left with the task of determining the *magnitude* of the change required in each disturbance, and the *probability* corresponding to the particular combination of worst-case s

We assume that we have a subset of n_{wc} disturbances, which we are including in the worst-case analysis, and that for each disturbance D_j we have a worst-case direction u_j defined as:

$u_j = 1$ if increasing D_j causes P_i to become worse

$u_j = -1$ if decreasing D_j causes P_i to become worse

Furthermore, we assume that we have identified the joint probability density function (*JPDF*) of the disturbances by ***tuning*** FABRICS to a particular IC manufacturing process. The JPDF of process disturbances is represented by the means and standard deviations of a hierarchical gaussian distributions, with the multiple levels corresponding to natural divisions in the process, i.e., intra-chip, inter-chip, inter-wafer and inter-lot. If we restrict ourselves to only one level, then each disturbance D_j can be represented by a single gaussian distribution with mean μ_j and standard deviation σ_j.

4.2.1.1. Algorithm for Worst-Case Analysis

In worst-case analysis, we are interested in determining the worst-case performance of a design, given a set of process disturbances modeling a particular manufacturing line. The tasks involved in worst-case analysis, namely the determination of a subset of disturbances, as well as the magnitude and direction of an adjustment in each of these disturbances to simulate the worst-case for each performance are illustrated in the following algorithm:

```
simulate the nominal performances of the circuit P

for each disturbance D_j

    perform perturbation analysis varying D_j

    for each performance P_i

        calculate the sensitivity of P_i to D_j

for each performance P_i

    adjust disturbances in the worst-case direction for P_i

    simulate worst-case P_i^wc
```

4.2.2. A Software Package for Worst-Case Analysis

WORCAN is a package of programs that performs worst-case analysis within the FABRICS-II/SPICE simulation framework. We now describe briefly each of the components of WORCAN.

Given the layout of the circuit, and a description of the manufacturing process, by performing process/device and circuit simulation, WORCAN produces voltage and current waveforms, from which the desired circuit performances can be extracted.

In this subsection we describe the extraction of timing, power and other performances from simulated waveforms. The tools implemented for this purpose are:

extract this program reads SPICE output files and generates binary data files for use by the performance extraction package, as well as by a variety of plotting programs.

wed is a *waveform editor* that allows the user to interactively (or by a command file mechanism) specify a set of measurements to be made on any or all waveforms. Such measurements may be any of:

1. Time(s) corresponding to a waveform crossing a given threshold, e.g., times when a voltage waveform crosses 2.5 V.

2. Value of waveform at a given time.

3. Peak(s) of a waveform, e.g., peak supply current.

4. Average of a waveform, computed as the integral of the waveform divided by the total time.

fed is a *performance editor* that reads **wed** output files and performs operations, interactively or through a command file, to calculate performances based on the measurements made on the waveforms, e.g., rise time can be calculated as the difference between the waveform crossing two prescribed thresholds. **fed** also allows the user to calculate arbitrary mathematical expressions with wed measurements as variables.

The program **rng** acts as a *supervisor* for worst-case analysis. It prompts the user for a set of disturbances to be included in the worst-case analysis, perturbs each in turn and extracts the performances for each value. A table of performances corresponding to the perturbed values of each disturbance is produced. Using this table, rng will check the monotonicity of each performance/disturbance relationship, and output warnings for cases where the worst-case situation is not well defined. **rng** will also produce a matrix of sensitivities and worst-case directions for all performances, and will simulate the worst-case for each performance. Fig. 4-16 shows an overview of WORCAN.

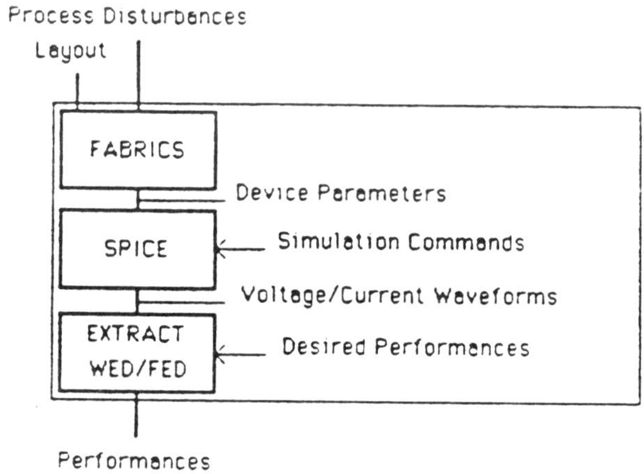

Figure 4-16: WORCAN system.

4.2.3. Examples

In this subsection we present three examples of the application of this methodology, and compare the results against those generated by the traditional method of performing worst-case analysis with respect to device parameters, and by Monte Carlo methods.

A major concern in any such approach is the computational cost required, as well as the accuracy of the results obtained. By far the largest component, typically greater than 90%, of the overall computational cost in the application of any worst-case analysis methodology, is the cost of circuit simulation. Two approaches can be used to minimize this cost:

- Simulate only *critical* portions of the circuit, such as a critical path in cases where delays are critical.

- Find the set of process disturbances that correspond to the worst-case performance of a small sub-circuit, such as an individual gate, and use that set to simulate the worst-case performance of the overall design. This eliminates the cost of finding the *sensitivities* of individual design performances, but relies on the assumption that *generic* worst-case files (e.g. increased power dissipation, or decreased speed) can be generated and will be valid across a range of designs. This is similar to some of the worst-case approaches which use a variety of device parameter sets, such as fast, slow, high-power, to generate worst-case performances. For cases where the worst-case performances depend on the *matching* of devices, however, worst-case parameters must be defined at two levels, such as intra-chip and inter-chip.

Inverter Chain.

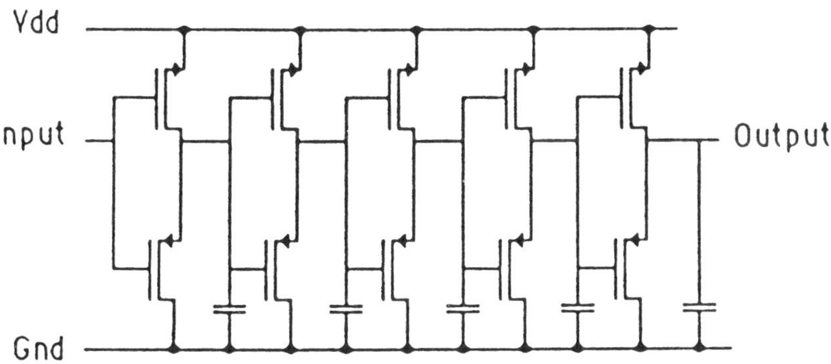

Figure 4-17: Inverter chain.

Fig. 4-17 shows a chain of 5 CMOS inverters with loads on each output. The performances of interest were the *total delay* and the power dissipation as the input to the chain is pulsed. The computational costs of the various phases of performances evaluations, on a VAX-11/785 running UNIX 4.2 bsd, were:

1. Process/Device simulation (FABRICS-II): 1.1 sec.

2. Circuit simulation (SPICE 2G7): 39.5 sec.

3. Performance extraction (extract/wed/fed): 0.7 sec.

For comparison, the worst-case performances were calculated by three methods:

- With respect to *device* parameters, i.e., we applied a similar methodology of finding sensitivities of the performances to each device parameter , determining a worst-case direction, and finally simulating a worst-case situation.

- With respect to process disturbances , as outlined above.
- Using Monte Carlo analysis.

For this example, we chose the 2σ points of both the process disturbances, as well as the device parameters , as the cutoff point for the analysis. The 2σ points for device parameters were generated by running FABRICS for a large number (250) of devices, and extracting the distributions of all device parameters . Fig. 4-18 shows a histogram of the threshold voltage of a p-channel device, showing the mean and the $\pm\ 2\sigma$ points.

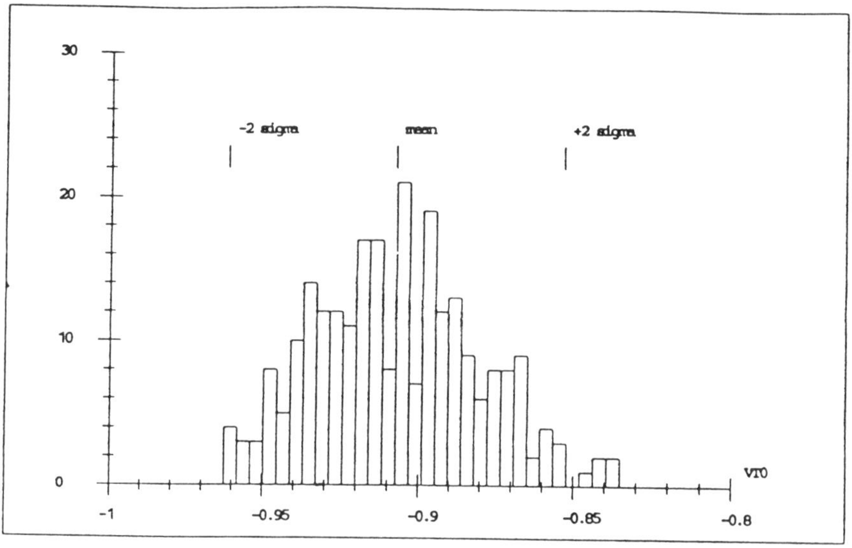

Figure 4-18: Histogram of p-channel thereshold voltage.

The device parameters used in this worst-case analysis were the lengths, widths and the intrinsic transconductances of both the N and P type devices. Table 4-3 shows the results of simulating variations of $\pm\ 2\sigma$ in the chosen device parameters on the performances[3], which were the total delay,

[3] In the actual simulations both the $\pm\ 1\sigma$ and the $\pm\ 2\sigma$ were simulated.

rise time of the last node in the chain, and total power dissipation. The worst-cases , for delay and power dissipation, were simulated by adjusting all device parameters 2σ in the worst-case direction, and are also shown in the table. Since device parameters are not independent, this analysis results in an unrealistic combination of device parameters.

	total delay (ns)	rise time (ns)	total power (mW)	WC-delay	WC-power
N-KP +	6.552	1.3145	0.3269		
NOMINAL	7.059	1.4228	0.3282		
N-KP -	7.691	1.61	0.3318	*	*
P-KP +	6.634	1.3703	0.3215		
NOMINAL	7.059	1.4228	0.3282		
P-KP -	7.466	1.539	0.3245	*	*
L-P +	8.81	1.613	0.3584	*	*
NOMINAL	7.059	1.4228	0.3282		
L-P -	5.5451	1.2839	0.3160		
L-N +	8.055	1.624	0.3489	*	*
NOMINAL	7.059	1.4228	0.3282		
L-N -	6.0655	1.2592	0.3096		
W-N +	6.798	1.3045	0.3704		*
NOMINAL	7.059	1.4228	0.3282		
W-N -	7.717	1.708	0.2897	*	
Worst case for delay	12.35	2.291	0.3017		
Worst case for power	11.02	1.874	0.4430		

Table 4-3: Performance of inverter chain versus device parameters.

We performed a worst-case analysis with respect to process disturbances, and Table 4-4 shows the results of simulating variations of $\pm\ 2\sigma$ in the line-width variations in the nitride and polysilicon layers, the diffusivity of boron, the linear dry oxidation rate, and the substrate concentration. Again, the worst-case performances are shown.

For the Monte Carlo analysis, we used FABRICS-II and SPICE to generate 250 samples of the performances. Two analyses were performed, in the first, we considered only geometric process disturbances, i.e., photolithographic line-width variations and misalignments; while in the second analysis, all the process disturbances were allowed to vary. Fig. 4-19 and Fig. 4-20 show histograms of the power dissipation in the inverter chain corresponding to the first and second analyses, respectively. It is clear from the figures that the variations in dimensions constitute only a part of the overall variation in the performances.

Table 4-5 shows a comparison of the worst-cases generated by the three

	total delay (ns)	rise time (ns)	total power (mW)	WC-delay	WC-power
linewidth nitride +	7.141	1.426	0.3116	•	
NOMINAL	7.059	1.4228	0.3282		
linewidth nitride -	6.885	1.4131	0.3400		•
linewidth poly +	6.3948	1.4147	0.3200		•
NOMINAL	7.059	1.4228	0.3282		
linewidth poly -	7.778	1.541	0.3440	•	
diffusivity boron +	7.151	1.426	0.3286	•	
NOMINAL	7.059	1.4228	0.3282		
diffusivity boron -	6.799	1.4269	0.3324		•
linear-dry oxidation +	7.412	1.628	0.2969	•	
NOMINAL	7.059	1.4228	0.3282		
linear-dry oxidation -	6.672	1.3291	0.3809		•
conc. in substrate +	6.953	1.4463	0.3257		
NOMINAL	7.059	1.4228	0.3282		•
conc. in substrate -	7.118	1.427	0.3276	•	
Worst case for delay	8.459	1.738	0.2948		
Worst case for power	5.67	1.2098	0.3791		

Table 4-4: Performance of inverter chain versus process disturbances.

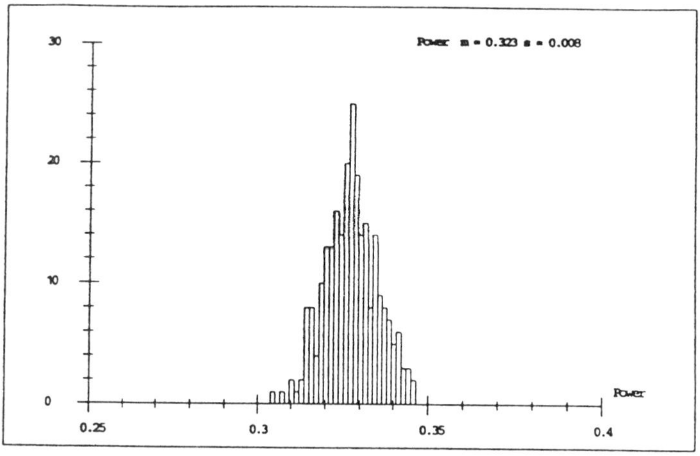

Figure 4-19: Power dissipation for first Monte Carlo analysis.

methods described above. We first show the nominal performances, then the worst cases for the delay and power generated with respect to device parameters and disturbances. Note that the worst-case performances generated with respect to device parameters are significantly more pessimistic than those generated with respect to process disturbances.

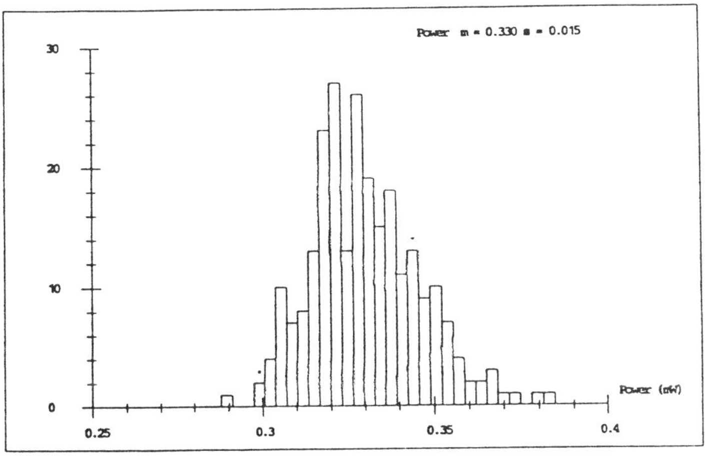

Figure 4-20: Power dissipation for second Monte Carlo analysis.

	total delay (ns)	total power (mW)
NOMINAL	7.059	0.328
Device Parameters		
Worst case for delay	(12.35)	0.301
Worst case for power	11.02	(0.443)
Disturbances		
Worst case for delay	(8.459)	0.295
Worst case for power	5.67	(0.379)
Monte-Carlo Analysis with geometric disturbances only		
mean	7.041	0.323
sigma	0.262	0.008
5.7 sigma point delay	(8.534)	
5.7 sigma point power		(0.369)
Monte-Carlo Analysis with all disturbances		
mean	7.041	0.330
sigma	0.314	0.015
5.7 sigma point delay	(8.831)	
5.7 sigma point power		(0.412)

Table 4-5: Worst-Case performances of inverter chain.

CMOS VCO.

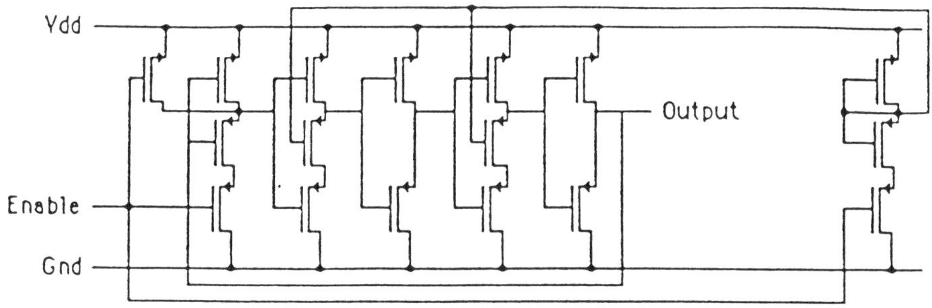

Figure 4-21: Circuit for CMOS VCO.

Fig. 4-21 shows a simple implementation of a CMOS voltage controlled oscillator. In this example, we calculated the *worst-case frequency,* assuming we wanted to maximize frequency, and using the first two methods illustrated in the previous example, namely with respect to device parameters and process disturbances.

The nominal frequency of oscillation, with the reference voltage derived from an inverter with its input and output tied together, was **46.4 MHz**. Note that the frequency, for this circuit, depends both on DC (the reference voltage) and AC (delay times through the stages) performances.

The device parameters used in the analysis were the lengths and widths and the transconductances of both the N and P type devices. The worst-case, simulated by adjusting all device parameters 2σ in the worst-case direction, was **24.0 MHz**. We performed a worst-case analysis with respect to process disturbances,and the disturbances simulated were the line-width variations in the nitride and polysilicon layers, the diffusivity of boron, the linear and parabolic oxidation rates, and the substrate concentration. The worst-case frequency in this case was **29.8 MHz**.

NMOS RAM.

For this example, we simulated one storage cell of 16-bit x 64 RAM implemented in a 4 micron NMOS technology. The simulations included the address decoding circuitry, storage cell, bit line precharge, as well as the sense amplifier. The performances calculated were power dissipation, peak supply current, as well as read and write access times.

Again the analysis was performed with respect to both device parameters and process disturbances. For the device parameters, the worst-case device parameters were the threshold voltages, intrinsic transconductances, and bulk coefficients of both the enhancement and depletion devices. The worst case disturbances were the line-width variations in the nitride and polysilicon layers, the diffusivities of boron and arsenic, the linear dry oxidation rate, and the substrate concentration.

For this analysis, worst-case parameters were shifted by 1 σ. The results of this worst case analysis are presented in Table 4-6. It can be seen that the worst-case calculated with respect to device parameters is significantly more pessimistic than that calculated with respect to the process disturbances.

	power (mW)	Ipeak (mA)	Twrite (ns)	Tread (ns)
nominal	1.45	1.1	16.9	23.8
worst case for each performance with respect to process disturbances				
power	2.28	1.8	13.6	15.7
Twrite	1.17	0.85	20.8	28.6
Tread	1.17	0.85	20.8	28.6
worst case for each performance with respect to device parameters				
power	2.30	1.9	12.7	25.5
Twrite	0.76	0.53	27.0	28.0
Tread	0.95	0.70	17.5	31.0

Table 4-6: Results of worst-case analysis for NMOS RAM.

4.3. Optimal Device and Cell Design Using FABRICS[4]

The complexity of VLSI circuits is such that more stringent restrictions are being imposed on the allowable variations in performance of semiconductor devices and subcircuits. However, at the same time, the performance of scaled down devices employed in VLSI circuits is more susceptible to the unavoidable random fluctuations which occur in the manufacturing process. In order to achieve satisfactory production yield, both process and device designs have to be optimized with these variations in mind.

With current *top-down* design methodologies, where functional blocks or *cells* are used in hierarchical fashion to assemble a design, it is possible to optimize individual cells to comply with constraints propagated from the higher design levels, e.g. constraints on the delay through some group of functional blocks, or the area of a cell. These constraints must be met by modifying the *geometry* of the design, i.e., it is assumed that the *topology* of the cells is fixed, but that some *layout parameters* (e.g. ratios of inverters) may be varied. Thus the concept of a *parameterized cell*, which would be a cell of fixed topology, with some variable geometrical features. Another group of parameters that have been traditionally used for optimizing the design are the model parameters of the IC components. However, since these parameters are in reality statistically dependent, the parameters controlling the fabrication steps must be used as *designable parameters* [72]. Clearly, the process parameters must be common for the entire circuit while the layout parameters can be set for each cell separately.

In this section we restrict our considerations to the optimization of the performance of a single cell. The optimization problem can thus be stated as finding the set of *designable variables* (process and layout parameters) which minimize some objective function (e.g. total power dissipated, delay time), while some constraints are placed on cell performances (e.g. noise margin) and geometry (cell area, aspect ratio). Since the performances are random variables, this is a statistical optimization problem.

---

[4]This section is based on the paper "Optimal Design of VLSI Minicells Using a Statistical Process Simulator" by A. J. Strojwas, S. R. Nassif and S. W. Director, In Proc. of 1983 ISCAS, May, 1983

4.3.1. Proposed Methodology

The probability density function (*pdf*) of cell performances depends on the designable parameters and *process disturbances* which represent random fluctuations inherent in the manufacturing process. These fluctuations cause small variations in device parameters within a cell (intra-die variations) and, more significant, inter-die variations ; which, combined, determine the parametric yield value. To solve the above statistical optimization problem, we have to determine the dependence of the *pdf*'s describing cell performances on the design variables, with fixed *pdf*'s for process disturbances. Since this dependence is not explicitly known, the problem involves a Monte Carlo simulation. Samples of device model parameters must be generated for each cell and then a circuit simulation (e.g. using SPICE) has to be performed. This process has to be repeated for a number of dies to obtain a sample of cell performances sufficiently large to estimate the *pdf*.

The key to the successful solution of the above problem is the availability of a *statistical process simulator*, such as FABRICS II. Due to its computational efficiency, the evaluation of the objective function and the constraints becomes a feasible task. However, this approach would lead to a prohibitive computational cost (due mainly to circuit simulation), especially if the optimization algorithm requires the evaluation of gradients of the objective function and the constraints with respect to the designable variables. Therefore, we propose another approach, in which a statistical optimization problem is transformed into an equivalent deterministic one.

Consider now the *pdf* of some cell performance. Since this *pdf* changes with changes in the designable variables, we have to re-estimate the distribution at each iteration in the optimization process. However, if we make the assumption that the *type of the pdf* remains the same, it is sufficient to extract the dependence of the *moments* of the *pdf* on the design variables. Specifically, we can establish a nonlinear regression model relating moments of the *pdf* of a given performance (e.g. mean value and standard deviation of the delay time) to the design variables.

To build the regression model, an efficient *plan of experiment* can be designed in which samples of the cell performances are evaluated for a small number of combinations of designable parameters. Performance evaluation is performed via process and circuit simulations. Moments of *pdf*'s are estimated for each sample corresponding to a set of designable parameters.

Then the moments are used to build a *least-squares* regression model. Once the model is established, it can be used for efficient optimization of an arbitrary function of circuit performances, thus the evaluation of the objective function, constraints and their gradients becomes very inexpensive.

4.3.2. Description of Experiment

To illustrate the methodology presented above, we now describe an experiment where the proposed approach was applied to an NMOS inverter.

The designable parameters considered were:

- ratio of pullup transistor length to pulldown transistor width, X_1
- minimum dimension, X_2
- gate oxidation time, X_3
- depletion threshold implant dose, X_4.

The performances that were considered were the power dissipation and the rise and fall times of the inverter. The power was taken as the average power over one transition cycle, while the transition times were estimated between 10% and 90% of the supply voltage.

FABRICS-II was *tuned* to an NMOS process, i.e., a set of *disturbances* was *identified* to fit FABRICS-II output to measured data from the process. The assumption is that these disturbances do not change as the process parameters are varied around the *nominal* operating point. FABRICS-II reads an input file containing the description of the circuit, filling *model-templates* (for this case, SPICE MOS transistor template) with element model parameters. An example of FABRICS-II input and output files is shown in Fig. 4-22, for a single inverter.

For the experiment, we considered a *wafer* consisting of 25 *test chips* each comprised of 12 inverters with various geometries. Fig. 4-23 shows the sequence of steps taken for each of the wafers. First FABRICS-II was used to generate the model parameters for each of the chips, then SPICE was run to produce circuit voltage waveforms, extraction routines were then used to compute the transition times and power dissipation.

The intervals over which the designable parameters were varied are given below, with the number of points taken over each interval:

Input file for FABRICS-II

```
* NMOS inverter
vdd 1 0 5.0
vin 2 0 pwl(0ns 0 10ns 0 20ns 5 50ns 5)
m10 1 3 3 0 d l=12 w=6
m11 3 2 4 0 e l=6  w=12
.tran 1ns 50ns
.print tran v(4) i(vdd)
.end
```

Output file produced by FABRICS-II

```
* NMOS inverter
vdd 1 0 5.0
vin 2 0 pwl(0ns 0 10ns 0 20ns 5 50ns 5)
m10  1  3  3  0 d10 l=  1.20e-05 w=  6.00e-06
.model  d10 nmos
+vto=-2.695e+00   kp= 4.923e-05 gamma= 1.213e+00
+tox= 5.599e-06 nsub= 1.660e+16   nss= 1.932e+10
+ xj= 8.710e-05   ld= 3.834e-07    uo= 7.986e+02
+ js= 4.090e-11  cgd= 2.363e-12   cgs= 2.363e-12
+cgb= 4.856e-12  cbd= 2.052e-06   cbs= 2.052e-06
+phi=-7.196e-01  nfs= 1.091e+10 ucrit= 6.241e+04
m11  3  2  4  0 e11 l= 6.00e-06 w= 1.20e-05
.model  e11 nmos
+vto= 1.020e+00   kp= 5.893e-05 gamma= 3.227e-01
+tox= 5.791e-06 nsub= 1.099e+15   nss= 1.914e+10
+ xj= 9.103e-05   ld= 4.775e-07    uo= 9.889e+02
+ js= 4.330e-11  cgd= 2.846e-12   cgs= 2.846e-12
+cgb= 4.781e-12  cbd= 2.015e-06   cbs= 2.015e-06
+phi= 5.792e-01  nfs= 1.086e+10 ucrit= 3.165e+04
.tran 1ns 50ns
.print tran v(4) i(vdd)
.end
```

Figure 4-22: Example of FABRICS-II input and output files.

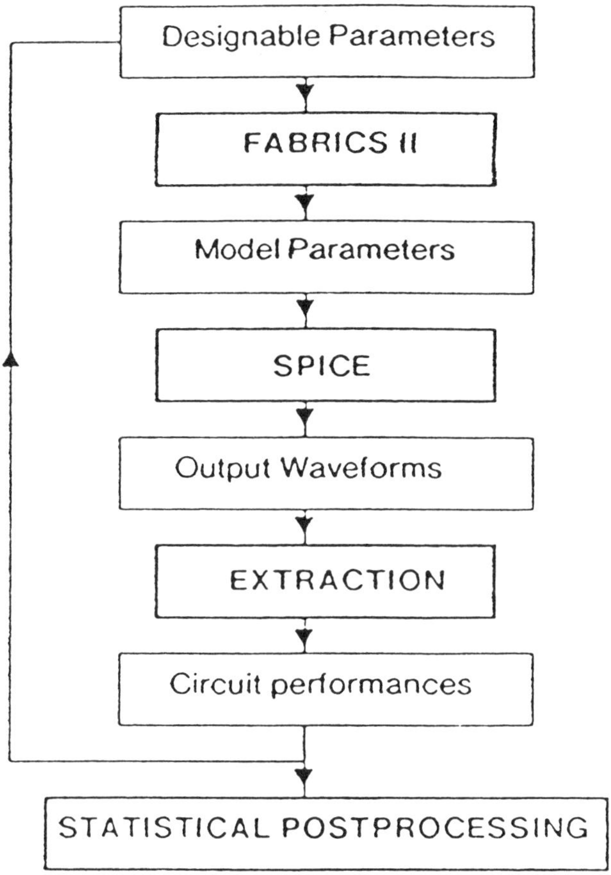

Figure 4-23: Computation flow for the experiment.

oxidation time (sec)	1000 to 1750	4 values
threshold dose (cm^{-2})	1.5E12 to 1.9E12	3 values
minimum dimension (μm)	2 to 5	4 values
pullup/pulldown ratio	3 to 5	3 values

4.3.3. Building the Regression Model

Statistical post-processing was used to produce the mean and variance of the performances for the inverter under all possible combinations of the designable parameters. Since the parameters of concern varied over many orders of magnitude (e.g. threshold dose and minimum dimension), the data was *normalized* with respect to the mean value of each sample.

Based upon the post-processed data, we built a non-linear regression model relating the moments of the probability distribution of the performances to the designable parameters. A least-squares regression algorithm was used, where the regression model was assumed to be a second order polynomial.

The results of the regression are shown in Table 4-7. The fit was evaluated using the multivariate correlation coefficient ρ [54]. It was found that, while models for power and rise time exhibit a good fit, the model for fall time has a lower correlation. This can be explained by the smaller dependence of the fall time, which is usually much smaller than the rise time, on the model parameters of the transistors.

$\mu(P)$	mean value of power dissipation	$\rho = 0.99$
$\sigma(P)$	standard deviation of power dissipation	$\rho = 0.70$
$\mu(\tau_r)$	mean value of rise time	$\rho = 0.99$
$\sigma(\tau_r)$	standard deviation of rise time	$\rho = 0.50$
$\mu(\tau_f)$	mean value of fall time	$\rho = 0.80$
$\sigma(\tau_f)$	standard deviation of fall time	$\rho = 0.66$

Table 4-7: Correlation coefficients for performances.

The expressions obtained for performances with the highest correlations are shown below:

$$\mu(P) = 6.6\text{-}3.25X_1\text{-}.58X_2\text{-}5.5X_3\text{-}.69X_4+.92X_1^2 \\ +.15X_2^2+2.2X_3^2+.383X_4^2+.2X_1X_2+.5X_1X_3 \tag{4.3}$$

$$\mu(\tau_r) = .29X_1+.875X_3\text{-}.655X_4\text{-}.674X_3^2 \\ +.3X_4^2+.735X_1X_3+.188X_2X_3 \tag{4.4}$$

From the expressions above, we found that the sensitivity of performances to the designable parameters was indeed as expected (e.g. strong dependence of rise time on gate oxidation and ratio).

4.3.4. Performance Optimization

In the final step, an implementation of the Han-Powell optimization routine [92] was used to optimize the performance of the inverter, using the macro-models built in the previous steps. In particular, we optimized the product of power dissipation and rise time, subject to constraints on area, power and rise time, as well as *box constraints* on the designable parameters. This is illustrated below, where we state the optimization problem, and show the optimums found, as well as the constraints applied:

minimize $\mu(\mathrm{P})\ \mu(\tau_r)$

subject to: (note that all values are normalized with respect to their mean values)

box constraints on designable parameters

$$0.75 < X_1 < 2.0$$
$$0.75 < X_2$$
$$0.75 < X_3 < 1.5$$
$$0.8 < X_4 < 1.2$$

cell area : $X_2^2(4+X_1) < 24$
power : $\mu(\mathrm{P})+\sigma(\mathrm{P}) < 1$
delay : $\sigma(\tau_r)/\mu(\tau_r) < 0.1$

The solution was obtained in 5 iterations (6 objective function evaluations); the objective function attains a minimum value of 0.94 for the following values of the designable parameters :

$X_1 = 0.873$
$X_2 = 0.75$
$X_3 = 1.17$
$X_4 = 0.96$

Note that the optimum value of the minimum dimension was *at* the constraint specified. This was caused by the fact that the lithography parameters in FABRICS-II (controlling line-widths and misalignments) were relatively small compared to the dimensions of the devices, and so had little effect on the variations in performance.

The example was run on a VAX-780 with floating-point-accelerator, under UNIX. The table below illustrates some of the computation times.

FABRICS-II	2.7 CPU seconds per chip
SPICE	32 CPU seconds per chip
REGRESSION	2.9 CPU seconds per model
OPTIMIZATION	14 CPU seconds.

Clearly, most of the computational effort is devoted to the circuit simulation and thus with the availability of faster (multilevel) simulators the efficiency can be drastically increased.

Chapter 5
Functional Yield

5.1. Introduction

In Chapter 1 we observed that functional yield losses, i.e., yield losses that are found during functional testing, are primarily due to local process disturbances (also called spot defects , point defects or random defects). Models that relate functional yield losses to a statistical characterization of spot defects have been studied since the early 1960's. Such models were developed to predict manufacturing yield by using defect densities extracted from an existing product. These models proved successful [111] for memory designs where defect densities extracted from one memory design could be used to predict yield of another memory design. The accuracy of predicting yield for this case was satisfactory because the topologies of the two memories were usually similar.

Unfortunately, this approach to yield modeling is not practical for VLSI circuits other than memories i.e., circuits in which the diversity of layout styles is large. Consequently, the sensitivity of performance to spot defects varies from one design to another and therefore information about defects extracted from one design may be inadequate in predicting yield for another design. Thus we are motivated to consider a methodology for predicting functional yield losses by using a product independent characterization of spot defects as well as precise analysis of the behavior of the defective IC. Towards this end, several yield prediction methodologies have recently been developed [66, 111, 112, 125, 73, 31, 32]. In this chapter we describe these approaches.

We begin our discussion by considering defect characteristics in more detail. Three yield modeling techniques are then introduced. The first of these uses analytical formulas to describe the probability of the functional failure of an IC in which a spot defect has occurred. In this technique, the IC layout is represented by a "virtual layout" which is topologically equivalent to the original layout but is simpler to use in computations. The second approach employs a straight forward Monte Carlo technique . In this approach the original IC layout is disturbed by randomly placed regions of extra or missing material that represents defects. The defective layout is then analyzed to determine the electrical consequences of the occurrence of the defect. In the third technique, yield is computed by an analytical formula that describe the probability of fault occurrence in the regions that are prone to spot defects. This approach results in significant simplifications in yield computation with relatively small loss of accuracy.

5.2. Basic Characteristics of Spot Defects

A spot defect is a randomly occurring region of extra or missing material in a layer used to fabricate an IC.[1] In this chapter we assume that spot defects have a disk shape. Not every defect causes malfunction of a circuit. Malfunctions occur when defects cause faults, i.e. when there is an error in the functional behavior of a circuit. This distinction is important because yield loss only occurs when there are faults. It is also important to distinguish between contamination and defects. Contamination can cause defects. For instance, a particle of any material deposited on the surface of a wafer is contamination. Whether or not contamination causes a defect depends on its physical nature and when in the process it was deposited. In general, the relationships between contamination, defects and faults are complex.

To simplify the discussion in this chapter we will be primarily concerned with the relationship between defects and yield although in some cases we will deal with faults. Of particular interest are those defects that can cause changes in IC topology which in turn cause changes in the DC operating point. Examples of such defects are extra and missing material defects, oxide pinholes, and small areas with excessive junction leakage [110, 73].

---

[1] Actually, in our approach, any process imperfection that can be characterized as local disturbance (defect) in the actual artwork of an IC layer can be included. An example of such an imperfection is a pinhole in the gate oxide or a crystal dislocation that causes leakage currents in a p-n junction.

5.2.1. Defect Mechanisms

As was discussed in Chapter 1 there exist a large variety of causes for defects. For instance, extra and missing material defects are caused by dust particles on a mask or on the surface of the wafer. During photolithography steps these particles lead to unexposed photoresist areas, or resist pinholes resulting in unwanted material or unwanted etching of material on a layer. For this reason such defects are sometimes referred to as *photo* or *lithography* defects [108]. These defects may occur in any layer but they are especially likely to occur in the poly, diffusion, metal, and via layers. Oxide pinholes are formed where contamination, crystal defects, or nitride cracking has occurred. Junction leakage defects are caused by crystal defects or contamination of the crystal lattice at the diffusion junction.

In general, for yield prediction a number of important defect characteristics are needed. First and most important, is the density of the defects, or the average number of defects per unit of area. If density of defects is nonuniform than a description of the nonuniformity is needed as well. The density of defects may also vary between the layers of an IC. For instance, it is a commonly observed phenomenon that the density of defects in the second metal layer are much higher than for example in the field oxide layer. Consequently, for adequate yield prediction the densities of defects per layer should be known. Finally, in any performance oriented yield analysis it is necessary to know defect size. This is important because electrical manifestation of the defect cannot be determined unless the defect size is known. In the remainder of this section we investigate the characteristics of spot defects.

5.2.2. Defect Spatial Distribution

In some processes defects tend to cluster within wafers, between wafers within a lot and/or between lots, due to time and space-varying particle concentrations in the air and chemicals used for fabrication. The spatial distribution of such defects can be modeled by a negative binomial distribution [107]:

$$\Pr(x) = \binom{\alpha + x - 1}{\alpha - 1} \frac{(\lambda/\alpha)^x}{(1 + \lambda/\alpha)^{x+\alpha}} \tag{5.1}$$

where x is the number of defects per wafer, λ is the expected number of defects, and α is the clustering coefficient . When $\alpha \rightarrow \infty$, the negative binomial becomes the Poisson distribution .

One can also use a two-level negative binomial distribution to describe the number of defects on a wafer. The first-level distribution models the variance between lots, and the second level models the variance between wafers within a lot.

It has also been observed that there exists nonuniformities of defects within wafers. A radial distribution in defect density [131, 41, 50] has been reported by several researchers. Failures are usually much more common in areas adjacent to the wafer edge than in other parts of the wafer. This feature may be, however, process specific and there are reported cases [71] in which the maximum density of the defects was located in the center of the wafer.

Generally, simple statistics are not sufficient to describe the phenomenon of defect spatial distribution. To model this distribution correctly a hierarchical model was proposed [18]. This model considers the effects of globally nonuniform distribution at the wafer level, local clustering at the chip level, and uniform distribution at the cell level. We will now describe this model in more detail.

Chip Level

On the chip level, defects are assumed to be affected by clustering. Instead of using compound or negative binomial statistics, this effect can be modeled as shown in Fig. 5-1, where D is the number of defects on a chip, and A, A_c, and A_d, represent the area of chip, macrocell, and device, respectively.

In order to model clustering, it is assumed that defects land on a cell one after another. Let X be a random variable which denotes the number of defects occurring on the cell. Also let $Pr\{x|X = i\text{-}1\}$ be the conditional probability for the ith defect occurring on the cell which already has $(i\text{-}1)$ defects. Then it is assumed that:

- **Assumption 1:**
  $Pr\{x|X = 0\}$, the probability for the first defect is proportional to the cell area, A_c. In other words, we assume that uniform distribution rather than defect clustering prevails for the first

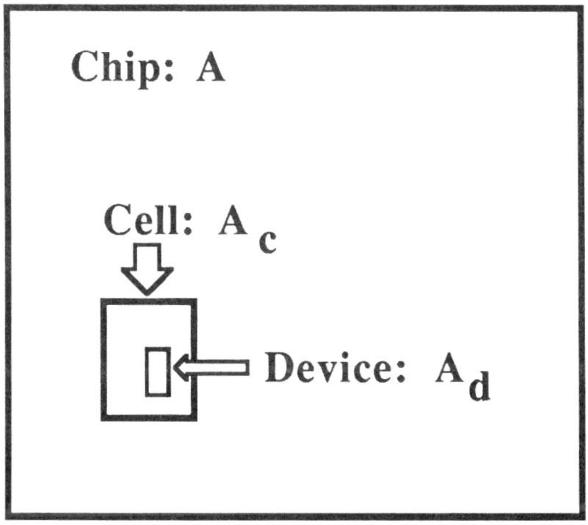

Figure 5-1: Hierarchical nature of IC layout.

defect.

- **Assumption 2:**
$Pr\{x|X = i\text{-}1\}$ is a monotonically increasing function of the random variable X, i.e. $Pr\{x|X = a\} < Pr\{x|X = b\}$, when $a < b$.

- **Assumption 3:**
$Pr\{x|X = \infty\}$ approaches 1 as X approaches ∞. Thus if there is already a large number of defects in the cell, the next defect will likely occur in the same cell due to very strong clustering.

Similar assumptions can be made for $Pr^*\{x|X = i\text{-}1\}$ which is the conditional probability for the ith defect landing within the area $(A\text{-}A_c)$ that already has $(i$ - 1 $)$ defects. We can write the following equations:

$$
\begin{aligned}
&P_1 = Pr\{x|X=0\} = \alpha A_c \\
&\ldots\ldots\ldots\ldots \\
&P_i = Pr\{x|X=i-1\} > P_{i-1} \qquad (5.2) \\
&\ldots\ldots\ldots\ldots \\
&P_\infty = 1
\end{aligned}
$$

and

$$
\begin{aligned}
&P_1^* = Pr^*\{x|X=0\} = \alpha(A-A_c)\\
&\dots\dots\dots\\
&P_i^* = Pr^*\{x|X=i-1\} > P_{i-1}^* \qquad (5.3)\\
&\dots\dots\dots\\
&P_\infty^* = 1
\end{aligned}
$$

where α is a proportionality constant, and P_i and P_i^* denote the conditional probabilities, $Pr\{x|X = i\text{-}1\}$ and $Pr^*\{x|X = i\text{-}1\}$, respectively. Note that the functions for P_i and P_i^* can be of any type as long as the above assumptions are satisfied. This means that the solutions for P_i and P_i^* may not be unique.

Here we assume that:

$$P_i = Pr\{x|X=i-1\} = 1-(1-P_1)exp\{-C(i-1)\} \qquad (5.4)$$

$$P_i^* = Pr^*\{x|X=i-1\} = 1-(1-P_1^*)exp\{-C(i-1)\} \qquad (5.5)$$

where C is a parameter characterizing the increase rate of P_i and P_i^* and, for simplicity, is assumed to have the same value in both equations.

Note that since the distribution of these D defects on A is not uniform, the number of defects occurring on A_c can be any number from 0 to D. Therefore to characterize clustering, the probabilities of all these possibilities must be known.

Consequently we can say that:

$$
\begin{aligned}
&Pr\{X=1\} = P_1\\
&\dots\dots\\
&Pr\{X=i\} = Pr\{X=i-1\}Pr\{x|X=i-1\} = P_1P_2 \dots P_i \qquad (5.6)\\
&\dots\dots\\
&Pr\{X=D\} = Pr\{X=D-1\}Pr\{x|X=D-1\} = P_1P_2 \dots P_D
\end{aligned}
$$

where $Pr\{X=i\}$ represents the joint probability for which i defects occur in area A_c. Similar formulae can be derived for $Pr^*\{X=i\}$.

Thus the total probability P_{ti} for i defects ($i \in (0, D)$) landing on A_c can be obtained by using formula from Table 5-1 [18] in which P_{ti} for different value of i are shown

No. of defects in A_c	Total probability
0	$P_{t0} = Pr^* \{X=D\}$
1	$P_{t1} = \binom{D}{1} Pr\{X=1\} Pr^* \{X=D-1\}$
2	$P_{t2} = \binom{D}{2} Pr\{X=2\} Pr^* \{X=D-2\}$
....	....
i	$P_{ti} = \binom{D}{i} Pr\{X=i\} Pr^* \{X=D-i\}$
....	....
D	$P_{tD} = Pr \; \{X=D\}$

Table 5-1: Total probability for i defects in A_c.

and

$$\binom{D}{i} = \frac{D!}{i!(D-i)!} \tag{5.7}$$

Note that P_{ti} is a function of P_i and P_i^* which, in turn, are functions of parameters α and C. Therefore, the clustering in this model, can be well characterized by two parameters: α and C. To determine values of these parameters we can use two constraints. The first one is that the number of defects landing on A_c can be any integer from 0 to D (including 0 and D). Thus we have

$$\sum_{i=0}^{D} P_{ti} = 1 \tag{5.8}$$

where

$$P_{ti} = P_{ti}(C, P_1, P_1^*) \tag{5.9}$$

Note that the sum of the total probability for i defects occurring on area $(A\text{-}A_c)$ is also equal to one. However, Eq. (5.9) is equivalent to equation Eq. (5.8) and cannot be treated as an independent constraint.

The second constraint can be obtained by normalizing the area A_c by the chip area A:

$$\sum_{i=0}^{D} iP_{ti} = \frac{DA_c}{A} \tag{5.10}$$

Generally, due to the complexity of the above equations, numerical methods must be used to solve for α and C.

After α and C are known, P_{ti} can be obtained from Eq. (5.9). The probability density function for the number of defects occurring in A_c can then be calculated from:

$$f_X(x) = \sum_{i=0}^{D} P_{ti}\, \delta(x-i) \tag{5.11}$$

where

$$\delta(x-i) = \begin{cases} 1 & \text{if } x = i \\ 0 & \text{otherwise} \end{cases} \tag{5.12}$$

Cell Level

At the cell level, defects are assumed to be uniformly distributed due to the fact that the area of interest is usually very small. Let A_d denote the area of interest, and p and q denote the probability of one defect occurring and not occurring on A_d, respectively. Then, based upon the assumption of uniform probability relative to area, p and q can be expressed as below:

$$p = \frac{A_d}{A_c}, \quad q = 1 - \frac{A_d}{A_c} \tag{5.13}$$

We use the binomial distribution to calculate the probability of i defects landing on A_d, $P_{d,i}$:

$$P_{d,i} = \binom{X}{i} p^i q^{X-i} \tag{5.14}$$

Since in most cases, A_d is very small as compared with A_c, the number of defects in A_d can be assumed to be either 0 or 1. Thus, for A_d we can write

$$P_{d,0} + P_{d,1} \simeq 1 \tag{5.15}$$

To find $P_{d,0}$, let

$$Y = \binom{X}{0} p^0 q^X = q^X \tag{5.16}$$

where Y is a random variable since it is a function of random variable X.

To make $P_{d,0}$ solvable, it is reasonable to approximate $P_{d,0}$ with $\overline{Y}$, i.e. the expected value of Y. Then, we have

$$\overline{Y} = P_{d,0} = \int_{-\infty}^{\infty} y \, f_Y(y) \, dy \tag{5.17}$$

where $f_Y(y)$ is the *pdf* of y and can be obtained from $f_X(x)$, based upon the theorem for random variable transformation. By applying this theorem one can show [19] that the probability for having 0 defects in A_d becomes

$$P_{d,0} = \sum_{i=0}^{D} \left(1 - \frac{A_d}{A_c}\right)^i P_{ti} \tag{5.18}$$

Finally, from Eq. (5.15), P_{d1}, can be found through

$$P_{d,1} = 1 - P_{d,0} \tag{5.19}$$

Note that in Eq. (5.18), if A_d/A_c is very small, then $\overline{Y}$ approaches 1. This means that for very small device area, A_d, the probability for having no defects in it, is approximately equal to 1.

Hence, the above model can be effectively used to model the nonuniform distribution of defect densities within the wafer. It is more flexible than other models and therefore should be useful not only to model typical defect clustering effects but also a large class of semi-regular nonuniformities such as for instance donut shape.

5.2.3. Distribution of Defect Radii

Most approaches to modeling spot defects can handle nonuniformities of defect density only and ignore completely the fluctuations that occur in the sizes of the defects. However, in a modern VLSI process, we can expect to see a dramatic increase in the number of "small" defects that in fact manifest themselves as faults. The importance of this effect have been illustrated by an experiment [72] in which it was shown that two IC designs with identical areas and strips of parallel metal lines had vastly different yields. In this particular experiment there was more than a twofold difference in yield. In both designs metal strips had identical pitch. The design with the lower yield had a 0.5 μm smaller spacing than the one with the higher yield. Of course, decreased spacing increase sensitivity to smaller defects. But such a difference in yield cannot be predicted unless defect size distribution is known and taken into account.

But accurate modeling of the defect size is not trivial. The actual radius of a defect is a function of both the radius of the contaminating particle that causes the defect and the lateral process deformations such as line width variations. In this section we show how to compute the probability density function associated with defect radius that takes into account both of these factors.

Estimation of the probability density function that describes the radius of a mask defect has typically been done through the use of test structures with narrow meander metal strips [112]. Experimental data collected from such test structures lead to the conclusion that this probability density function is of the form $1/x^3$ where x is a diameter of the defect (see e.g. [112]). Recent studies [71] suggest, however, that such a relationship is not universal and therefore a more sophisticated model is required. More specifically, a better probability density function is one based upon a combination of several nonsymmetrical density functions, each of which characterizes one of the mechanisms that causes spot defects. A set of weights that control the contributions of each function towards determining the final defect radius can then be found by fitting the postulated model to experimental data.

One of the density functions suitable for describing the distribution of the radius of a defect introduced by a single mechanism is the Rayleigh distribution [86]:

$$f_r(x,\alpha) = \frac{x}{\alpha^2} exp(-\frac{x^2}{2\alpha^2}) \quad \text{for } x > 0$$

and (5.20)

$$f_r(x,\alpha) = 0 \quad \text{for } x < 0$$

where α is the distribution parameter. It was found that in order to describe the experimental data reported in the literature, Eq. (5.20) can be used in the following formula:

$$f(R) = \{[\beta_1\, ry\,(R,m_1^*,\alpha_1) + (1-\beta_1)\, ry\,(R,m_2^*,\alpha_2)]\beta_2 + \\ + (1-\beta_2)\, ry\,(R,m_3^*,\alpha_3) \qquad (5.21)$$

where :

1. $f(R)$ is the probability density function describing the radii of the contamination or the particles;

2. $ry\,(R,m_i^*,\alpha_i) = f_r(R-m_i^*,\alpha_i)$

3. β_1,β_2 and m_i^* and α_i for $i = 1, 2, 3$ are parameters of $f(R)$.

In this formula, parameters m_1^*, m_2^*, α_1 and α_2 are used to characterize airborne and equipment generated particles; parameters m_3^* and α_3 are used to characterize layer defects caused by mechanical defects, and β_1 and β_2 are weights.

A simple fitting technique can be used to compute the parameters of Eq. (5.21) by minimizing the difference between the density function Eq. (5.21) and the measured histograms of the radius of defects on the surface of the mask.

The key to the successful application of Eq. (5.21) is an adequate defect radius measurement technique that assures reasonable accuracy and is inexpensive enough to be practical for large sample sizes. Such a technique that employs a Double-Bridge test structure has been proposed and applied for the extraction of the defect sizes in the interconnect of VLSI ICs [72] .

5.2.4. Distribution of Defect Radii Within Layer

From the above discussion it is clear that the number of defects that cause faults is a function of defect size. Since a significant number of the spot defects are introduced through lithography and etching steps, and each of these steps is performed similarly for the vast majority of processing steps, we would expect to have a similar number of defects within each IC layer. Experimental data shows that this is not the case. In this section we explain one of the reasons for differences in the effective defect densities in different IC layers. Towards this end we compute defect size distributions in a number of layers assuming that the distribution of contaminants that cause these defects are identical.

Consider a contaminating particle that is located on a mask, the surface of a photoresist layer, or on the surface of the wafer. Such a particle is engraved in an actual IC layer with process induced lateral deformations of various kinds. Thus, the actual radius of the defect, i.e. the radius of the defect in the IC structure, see Fig. 5-2, is described by a random variable which is the sum of the radius of the defect and any process induced lateral deformations or line registration errors. Such errors can be described by a Gaussian distribution. Hence the distribution of defect radii can be computed as a convolution of Eq. (5.21) with a Gaussian density function.

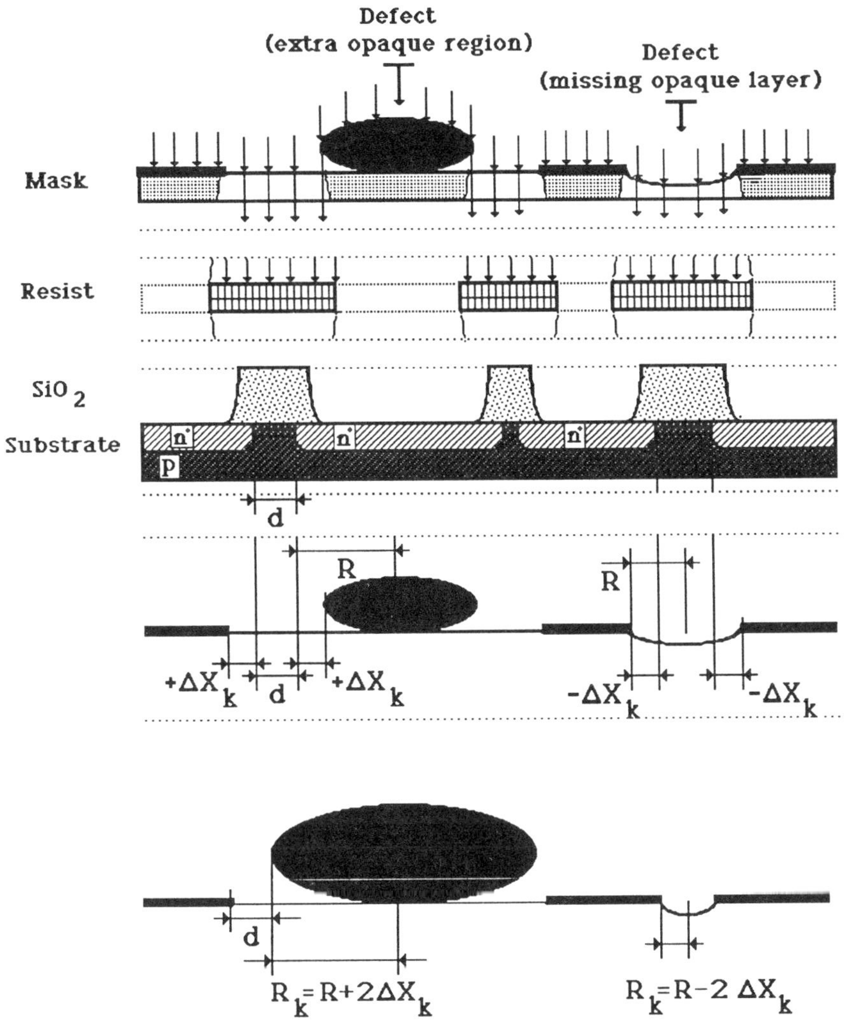

Figure 5-2: Relationship between radii of the defect on the mask and its image in the actual IC structure.

Note, that in order to determine whether a defect causes shortor break we do not need to know the defect radii itself, but rather we need to know the distances between the edges of the layout and the edge of the defect. In order to compute distances between the edge of the defect and edges in the layout, we assume that the locations of the layout edges are not disturbed by the line registration errors but that the defect radius is disturbed by double the line registration error. (See Fig. 5-2.) Thus, if the line registration errors for the k-th layer in the IC structure is described by the normally distributed random variable ΔX_k with a mean value μ_k and standard deviations σ_k, the radius of the defect in this layer, R_k, is given by

$$R_k = R + 2\Delta X_k \tag{5.22}$$

and , thus, the probability density function describing the distribution of R_k is:

$$f(R_k) = \int_{-\infty}^{+\infty} f(R) N(R_k - R, 2\mu_k, 2\sigma_k) dR \tag{5.23}$$

where $f(R)$ is given by Eq. (5.21) and $N(x,\mu,\sigma)$ describes the Gaussian distribution of a random variable x with mean and standard deviation μ and σ , respectively.

To evaluate differences in spot defect size distributions in various layers of a typical NMOS process we performed the following computations. The spot defects characteristics were assumed to be as reported in [112]. Then assuming typical values for line registration errors, we computed parameters of Eq. (5.21) to obtain the agreement between model and measurement data. Finally, we computed density functions which described defect radii in various layers for standard NMOS technology. The details of this computation, and the results obtained can be found in [69]. A summary of the results is shown in Fig. 5-3. (Distances which are negative indicate possible shorts or breaks in the layout being analyzed.)

As one can see there exist significant differences among density functions describing radii of the defect in various IC layers. This data also indicates that the higher the scale of the integration of the IC the more important these differences become.

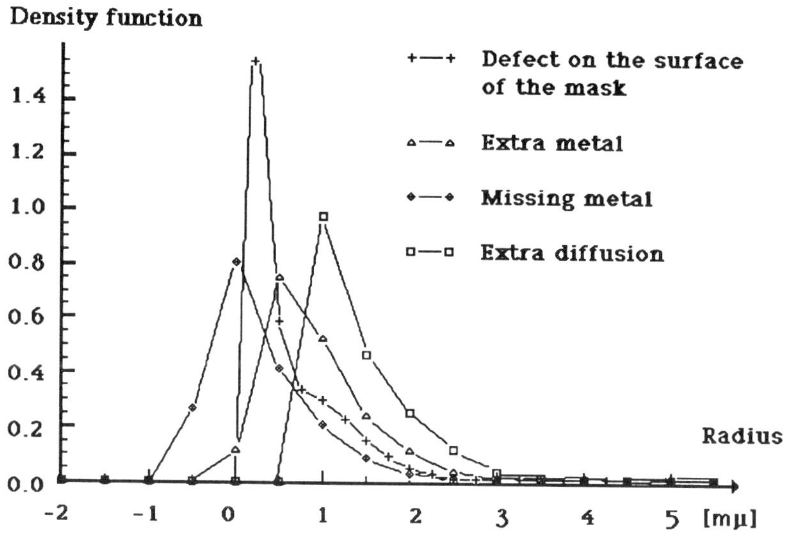

Figure 5-3: Distribution of defect sizes in various layers caused by the identical mask defets.

5.3. Yield Modeling Using Virtual Layout[2]

Conceptually predicting yield loss for an arbitrary layout and given defect size distribution is relatively simple. Note that in order to determine the probability of a defect that manifests itself as a fault, we must determine which fraction of defects cause permanent changes to circuit connectivity, i.e. a short or a break. To this fraction via a Monte Carlo approach, one simply places defects on the layout and then, by analyzing the area in the neighborhood of the defect, checks whether nonequipotential nodes are shorted or new nodes are created. If either of these situations occurs then a fault has occurred. The only difficulty in this approach is the sample size that has to be used -- it has to be very large if statistically valid results must be obtained. Note that the only computationally intensive operation is

[2]This section is based upon the paper:"Modeling of lithography related yield losses for CAD of VLSI circuits", by W. Maly, IEEE Trans. on CAD, Vol. CAD-4, No. 3, pp. 166-177, 1985.

the process of checking whether the defect of a given size and location "touches" (bridges or shorts) two regions or it breaks a conducting path. To see if this happens, the distances from the edge of the defect to the closest elements of the layout must be determined.

Note also that we can determine if a fault has occurred in a different way. By knowing the size of the defect one can determine all locations of the centers of a defect that cause a fault. The set of all such locations is called *critical area*. In the following subsection we describe an approach to yield estimation that is based on the concept of critical area.

5.3.1. Critical Area

As already indicated, spot defects can affect performance of an IC in two ways. They can break an equipotential region into two or more nonequipotential regions (see Fig. 5-4 a and b), or they can cause a short between two nonequipotential regions (see Fig. 5-5 a and b). Shorts and breaks are possible in both insulating and conducting layers of an IC.

Fig. 5-4 (c), Fig. 5-5 (c) and Fig. 5-6 illustrate the concept of critical area for a very simple pattern of conducting lines. Note that the width of the critical area, w_f, is proportional to the width of the path, w, and defect radius R_k (see Fig. 5-6). The width of the critical area w_f, and therefore the critical area $A_f(R_k,w)$ itself, are equal to zero when $R_k < w/2.0$. The critical area $A_f(R_k,w)$ increases (see Fig. 5-6 b) with slope $2L$ where L is a total length of the conducting path. The critical area has an upper limit equal to the total area of the region under consideration. Thus for a conducting path such as shown in Fig. 5-7, $A_f(R_k,w)$ saturates for $R_k = s/2+w$ (see Fig. 5-7 b).

However, in general, computation of the critical areais more complex because the topology of the actual IC is much more complicated than in the above example. Thus determining the critical area involves a sequence of layout resizing operations or a Monte Carlo simulation. For the present, let us assume that we are able to compute a critical area as a function of defect radii, for each layer of the IC structure. We will return to this point later. Let us discuss the yield loss formula assuming that $A_k(R_k)$ is known.

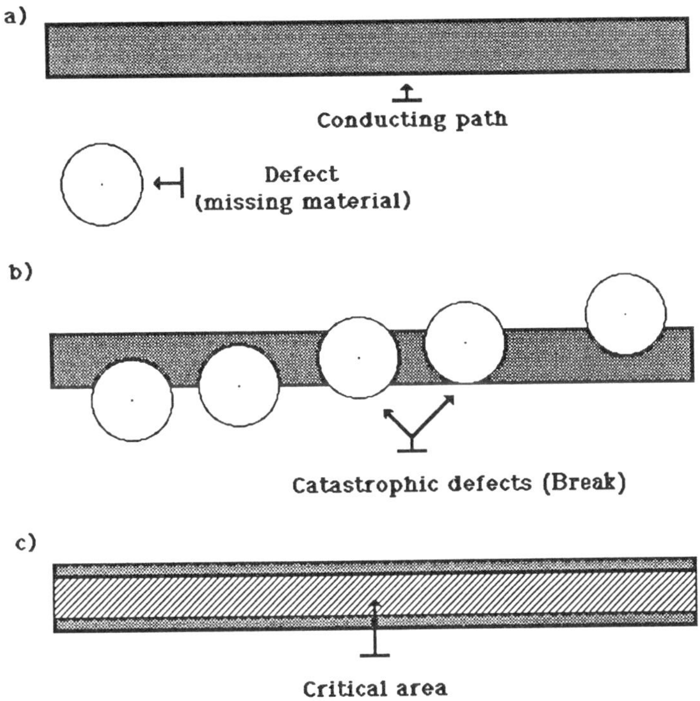

Figure 5-4: Break.

5.3.2. Spot Defect Related Yield Losses

The probability of the event that v_k spot defects with radius R do not cause a short (or break) in the k-th IC layer, Y_p^k, can be computed from the following Poisson formula: [86]

$$Y_p^k = exp\{-v_k p_k\} \tag{5.24}$$

where p_k is a probability that single defect causes a short (or break). But:

$$v_k p_k = \frac{v_k}{A_{ch}} A_k(R) = D_k A_k(R) \tag{5.25}$$

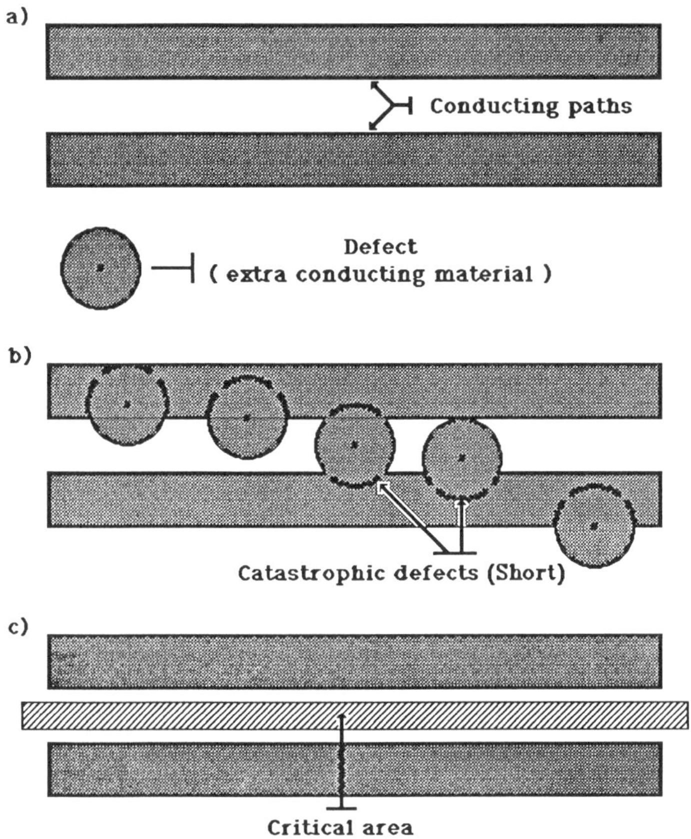

Figure 5-5: Short.

where D_k is the density of the defects, $A_k(R)$ is the critical area computed for defects having radius R and A_{ch} is the area of the chip. So that Eq. (5.24) takes the form:

$$Y_p^k = exp\{-v_k p_k\} = exp\{-D_k A_k(R)\} \tag{5.26}$$

Note now that Eq. (5.26) is still valid for the case in which the radius of a spot defect is described by a random variable **R**. Consider a sequence of segments $[R_k^i, R_k^{i+1}]$ for $i = 1, 2, 3, ..., N$ arranged in such a way so as to cover the majority of the actual values of **R**. For each of these segments we can estimate $p(R_k^i)$, the probability that **R** is contained in the segment

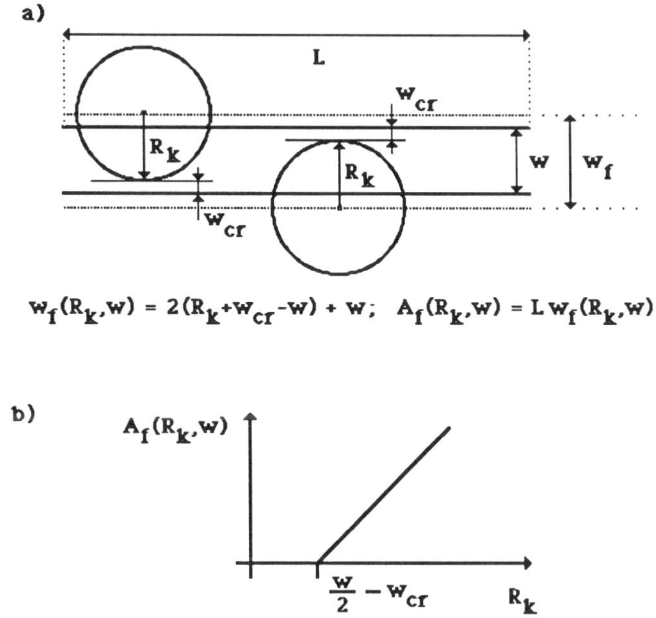

Figure 5-6: Critical area.

$[R_k^i, R_k^{i+1}]$. Thus the product, $D_k p(R_k^i)$ is the density of defects whose radii are in the segment $[R_k^i, R_k^{i+1}]$. If now we assume that

$$R_k^{*\,i} = (R_k^i + R_k^{i+1})/2.0 \tag{5.27}$$

and that $A_k^i(R_k^{*\,i})$ is the critical area computed for $R_k^{*\,i}$, then from Eq. (5.26) we know that the probability of the event that the defect whose radius is contained in the segment $[R_k^i, R_k^{i+1}]$ causes a short (or a break) is described by:

$$exp\{-A_k^i\,(R_k^{*\,i})\,p(R_k^i\,)\,D_k\}$$

Using Eq. (5.26) we can show that :

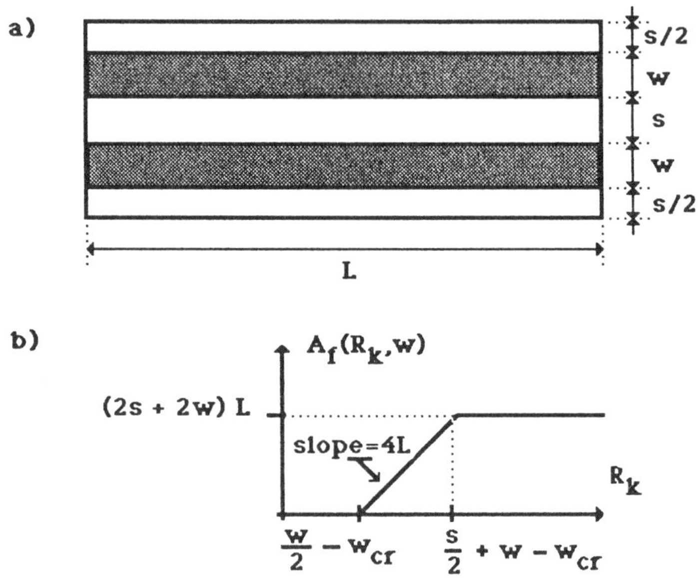

Figure 5-7: Critical area as a function of defect radius.

$$Y_p^k = \sum_{i=1}^{N} p(R_k^i)\, exp\{-D_k A_k^i (R_k^{*\,i})\, p(R_k^i)\} \tag{5.28}$$

where

$$p(R_k^i) = \int_{R_k^i}^{R_k^{i+1}} f(R_k) dR_k \tag{5.29}$$

and $f(R_k)$ is the probability density function given by Eq. (5.23).

Finally, note that if we assume that the IC under consideration is composed of K layers, then we must use Eq. (5.28) $2K$ times to determine the final yield Y_p:

$$Y_p = \prod_{k=1}^{2K} Y_p^k \tag{5.30}$$

5.3.3. Yield Losses Due to Lateral Process Deformations

Line registration errors themselves may cause shorts and breaks in an actual IC structure. To evaluate the probability of a short or break due to line registration errors, we have to find, for each nominal IC layer, the minimal distances representing the minimal width of the connection and the minimal spacing between connections in this layer. We call these distances *critical* distances and denote them by w_j and s_j, respectively. Knowing the probability density functions describing line registration errors for the j-th IC layer, one can compute the actual distances W_j and S_j. The actual distance is the nominal distance plus the random variable describing the line registration error. An IC will not contain shorts or breaks if all of the actual critical distances are greater than certain minimal technologically acceptable distances, w_{cr}^j, that are specified for each IC layer. Hence, in order to estimate the yield limited by the line registration errors, we have to find for each IC layer the probability of the event that both W_j and S_j are greater than w_{cr}^j. We denote this probability by Y_e^j. It can be shown, using simple geometrical considerations, that this probability equals the probability that the random variable $w_j+2\Delta X_k$ is greater than w_{cr}^j and it is smaller than $w_j+s_j-w_{cr}^j$ (See Fig. 5-8.) . Thus Y_e^j can be computed from the formula:

$$Y_e^j = \int_{w_{cr}^j}^{s_j+w_j-w_{cr}^j} N(x, 2\mu_j+w_j, 2\sigma_j)dx \tag{5.31}$$

where, as before , $N(x,\mu,\sigma)$ denotes the Gaussian distribution function whose mean and standard deviation are determined assuming that the line registration error ΔX_j has mean and standard deviation equal to μ_j and σ_j, respectively.

Hence, the yield limited by the line registration errors themselves is determined by the product:

$$Y_e = \prod_{j=1}^{K} Y_e^j \tag{5.32}$$

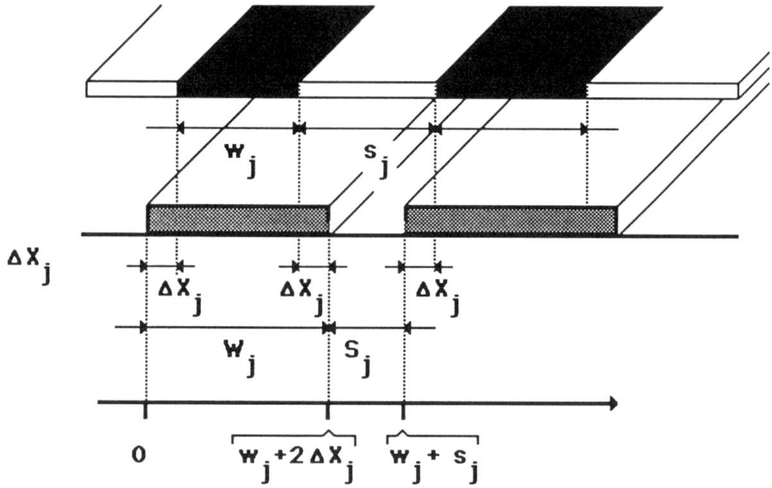

Figure 5-8: Distance between nonequipotential regions as a function of line registration error.

where K is the number of layers in the IC. An important feature of Eq. (5.32) is that it is useful when it is applied to regions bounded by edges belonging to the same category, i.e. those fabricated by the same mask. In practice, however, we must also consider shorts and/or breaks that occur between edges of different categories, e.g. between the edge of the polysilicon region and the edge of the diffusion region. In this case the end points of critical distances fluctuate independently and misalignment vectors can not be disregarded. It can be shown, however, that the actual value of a critical distance defined for edges fabricated by means of different masks can be easily expressed in terms of line registration errors and appropriate components of misalignment vectors. An example illustrating such expression is shown in Fig. 5-9 where s_{ij} and S_{ij} denote nominal and actual spacing between edges created by means of the j-th and i-th mask, respectively, are presented. If the random vectors ΔX_i and ΔX_j represent line registration errors, excluding misalignment effects, and ΔM_i and ΔM_j describe components of misalignment vectors for the i-th and j-th mask respectively then:

$$S_{ij}=s_{ij}+\Delta X_i+\Delta M_i+\Delta X_j+\Delta M_j \tag{5.33}$$

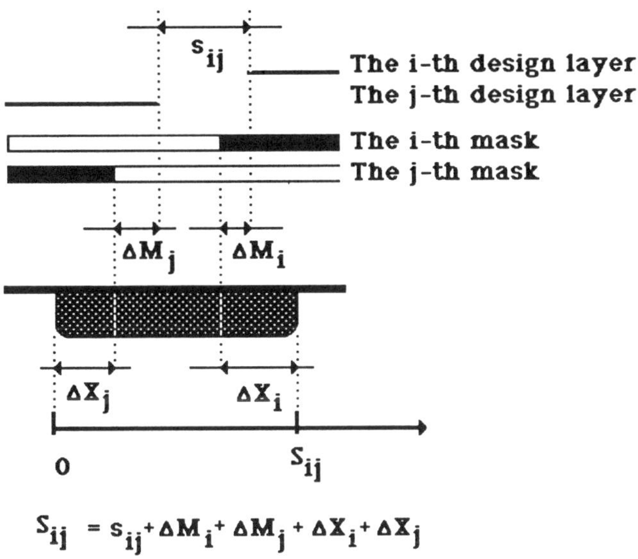

Figure 5-9: Spacing between edges of two regions as a function of misalignment and line registration errors.

and that the probability that the edges i and j are not touching one another is given by the integral:

$$Y_e^j = \int_{w_{cr}^j}^{+\infty} N(x,\mu,\sigma)dx \tag{5.34}$$

where μ and σ are computed from Eq. (5.33).

Hence, to include in the line registration error the mask misalignment effect we have to modify Eq. (5.32) as follows:

$$Y_e = \prod_{j=1}^{J} Y_e^j \tag{5.35}$$

where J is the number of all combinations of edges that may be shorted.

The Y_e^j are computed from Eq. (5.31) and Eq. (5.34) according to the categories of edges which are endpoints of the critical distance under consideration.

Eq. (5.35) is the probability of the event that the IC does not have shorts and breaks caused by line registration errors under the assumption that the structure does not contain spot defects. Hence, Eq. (5.35) describes a conditional probability. Eq. (5.30) describes the probability that the IC does not contain shorts (or breaks) which are due to spot defects and line registration errors combined together, thus it can be interpreted as the probability of the condition used to develop Eq. (5.35). Therefore, manufacturing yield can be expressed as the product of the probability that the IC does not contain shorts and breaks caused by spot defects and line registration errors and the probability of the event that the IC does not contain shorts and breaks caused by the line registration errors themselves, i.e.,

$$Y = \{\prod_{k=1}^{2K} Y_p^k\}\{\prod_{j=1}^{J} Y_e^j\} \tag{5.36}$$

where Y_p^k and Y_e^j are defined by Eq. (5.36) and Eq. (5.31) , respectively.

Finally, note that to find the critical distances in the nominal layout one has to perform exactly the same computation as is needed in geometrical design rule checking.

Further, observe that yield evaluation will usually be performed after geometrical design rules have been checked and any errors corrected. Therefore, instead of looking for edges which may be shorted due to line registration errors we may simply assume that these distances are such as specified by the design rules. Consequently, yield loss(defined by Eq. (5.35)) can be computed for all distances specified by the design rules without an analysis of the actual layout.

5.3.4. Critical Area Computation Using Virtual Layout

The yield model introduced above can be applied for practical purposes if one can express the critical area in terms of the defect size. Unfortunately, computation of the critical area as a function of defect size is not trivial. The difficulty lies in the large number of entities that have to be manipulated in order to determine the regions that are vulnerable to the defect. To simplify this task, we use the concept of "virtual layout", which is an artificial layout that is extracted from the nominal IC layout.

Before defining virtual layout observe that it is relatively easy [8] to statistically characterize a nominal IC layout in terms of such quantities as the total length of connections in a layer, the average spacing between connections, or the total area of active layer elements. It is also possible to compute histograms containing information about the fraction of the layout entities having a specific feature such as a certain width, etc. We may then use this data to build a layout that has the same statistical features as the nominal IC layout but arranged in the form of parallel strips. To illustrate this concept consider Fig. 5-10 in which a segment of the cell of a n-MOS circuit is shown; its metalization layer is highlighted in Fig. 5-11. Note that for this layer we can build a layout (see Fig. 5-12) which is vulnerable to spot defects in the same manner as the nominal IC layout, but for which it is possible to derive an analytical function that relates defect radius and critical area.

In general one can postulate a virtual layout that is an array of parallel conducting strips separated by insulating strips whose widths and spacings are specifiable parameters. For such a layout, it is easy to formulate an analytical function that relates the critical area to the defect radii and parameters of the layout. (An example of a virtual layout along with the formulas for computing the critical area, has already been presented in Fig. 5-6 and Fig. 5-7.)

Note that application of virtual layout can be straightforward. It involves extracting from the nominal IC layout the statistical information that characterizes the connection width, spacings, lengths, etc. Values of the virtual layout parameters can be chosen such that statistical characterization of the virtual layout is identical with the statistical characterization of the nominal IC layout. It is then possible to derive analytic expressions that relate the critical area of the virtual layout to the defect radii. These expressions should accurately estimate nominal IC layout. Of course, accuracy of this estimation is limited by differences in the detail of the nominal and virtual layout. For the case of a large circuit designed with consistent design style, locally significant details of the layout

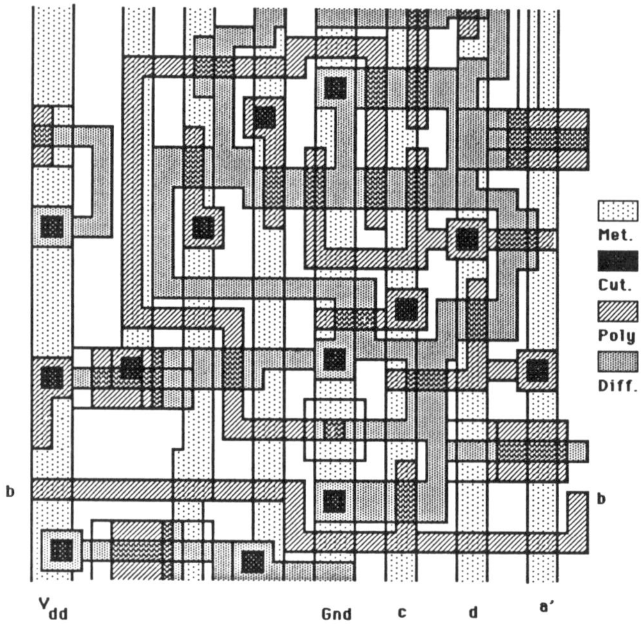

Figure 5-10: Segment of the NMOS layout.

are unimportant as far as the overall probabilities of shorts or breaks are concerned. The only important features of the layout are those which have statistical significance, i.e., those features which are represented by the data obtained from the statistical analysis extracted from the nominal IC layout.

5.3.5. Examples of Application of the Virtual Layout Method

As a first example of the application of the virtual layout method consider the optimization of design rules. Typically, design rules are expressed in terms of a common unity distance called lambda (λ). Optimal design rules result from determining the smallest value of λ for which the yield is acceptable.

The application of the virtual layout for design rule optimization is very straightforward because one can easily construct a virtual layout for each rule separately. Then critical area functions can be built and yield can be

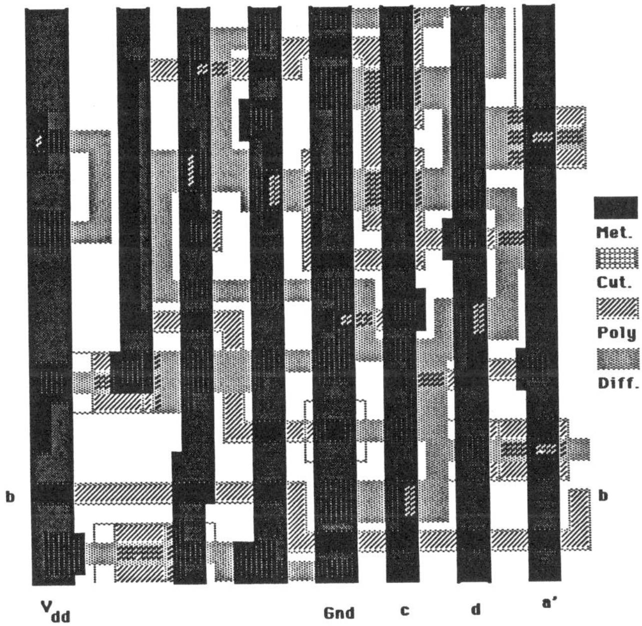

Figure 5-11: Metallization layer.

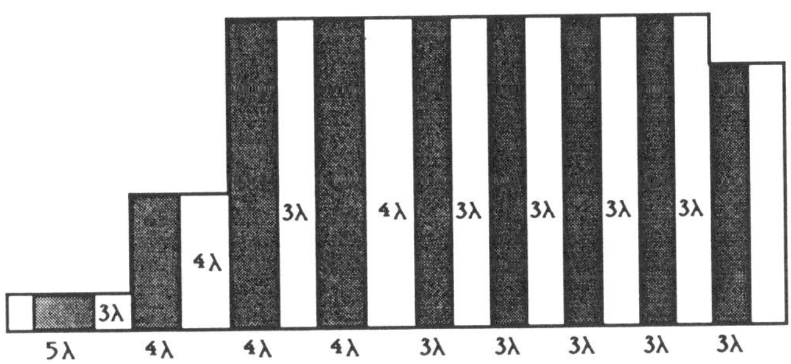

Figure 5-12: Virtual layout for the metalization layer shown in Fig. 5-11.

computed for a die of a given area. In [69] more precise description of the computations are presented. Fig. 5-13 summarizes results that have been obtained. Using this figure one can find for each of the critical distances (design rules) such distance which is minimal but at the same time assures large enough yield.

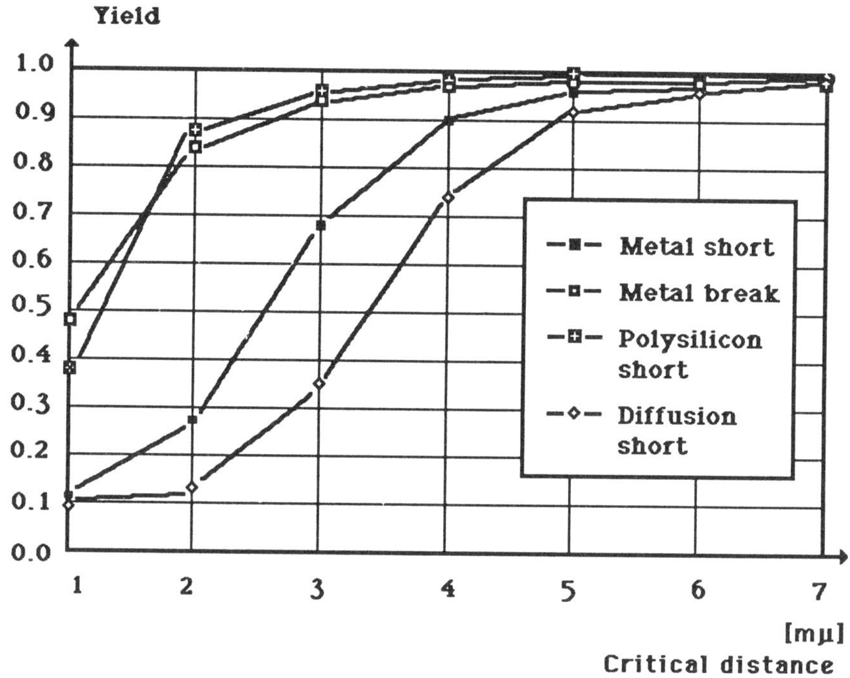

Figure 5-13: Yield as a function of critical distances (design rules).

The second illustration deals with the scaling (shrinking) of VLSI layout. Scaling is usually performed to assure a maximal number of fault-free IC die on a single wafer. An optimization of such a number was performed assuming a fixed size wafer and computing yield from the wafer as a function of scaling factor. In the computations it was assumed that the 1 cm^2 die, designed in a 5 μm NMOS technology must be scaled down and its virtual layout was created using statistical characteristics extracted from a typical VLSI student project described in [8]. The results of the computations taken from [69] are summarized in Fig. 5-14.

Note that there exists a sharp optimum for wafer yield indicating that optimization of the design rules, such as described above, may provide substantial economical gains. Note also that in both cases relatively complicated problems were solved via a rather simple means.

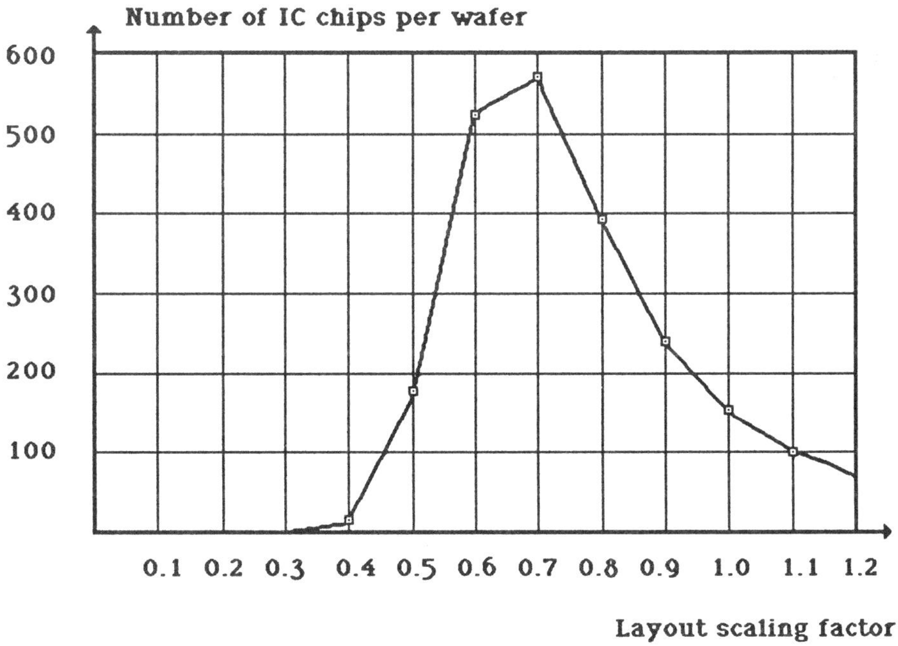

Figure 5-14: Number of IC dies as a function of a scaling factor.

5.4. Monte Carlo Approach to Functional Yield Prediction

An alternate means for determining the effects of catastrophic faults on yield is through the use of Monte Carlo simulation. In such a simulation, random number generators are used to generate defects and place them on the layout. Once the defects have been placed on the layout, a series of fault analysis procedures are used to determine what, if any, circuit faults have occurred. The ratio of the number of defects that do not cause faults to the total number of defects generated is interpreted as functional yield. This approach has been implemented in a program named VLASIC, that is described below.[3]

[3]The procedure described here deals with a standard NMOS process but it can easily be extended to other technologies.

5.4.1. The VLASIC Yield Simulator[4]

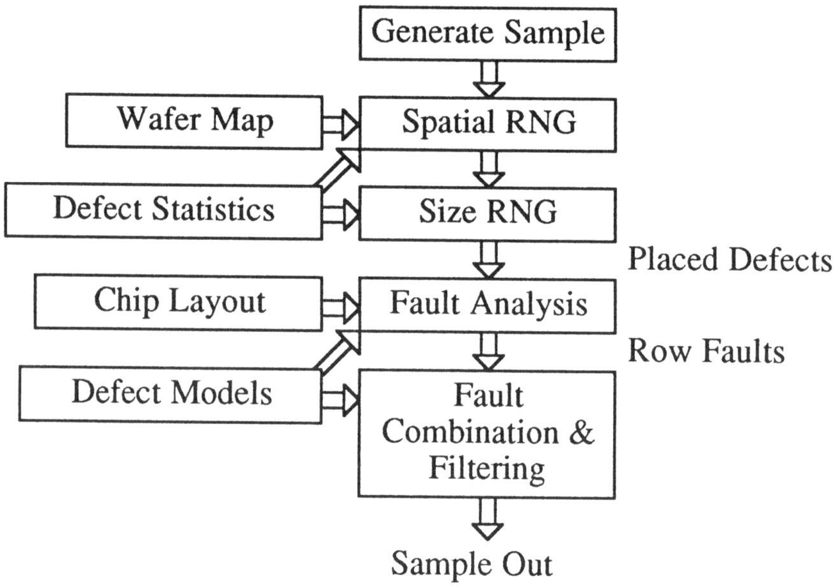

Figure 5-15: Structure of VLASIC.

The basic structure of VLASIC is shown in Fig. 5-15. Observe that the defect random number generators are organized in a hierarchical manner, with several generators for the spatial distribution, and one for the size distribution. These generators take as inputs parameters such as the mean and variance of the defect density. A wafer map is used in the process of placing defects within a wafer. The output of the random number

[4]This section is based upon the paper: "VLASIC: A catastrophic Fault Yield Simulator for Integrated Circuits", by D.H. Walker and S.W. Director IEEE Trans. on CAD, Vol. CAD-5, No. 4, pp. 541-556, 1986.

generators is a list of defect types, their location within the die sample, and defect diameters.

A series of fault analysis procedures, which are the key elements of VLASIC, are then called to examine the geometry of the layout within the neighborhood of the defect to determine if any circuit faults have occurred. These procedures use general-purpose polygon operations, avoiding the need to place any restrictions on chip layout (i.e., such as limiting acceptable layout styles to Manhattan geometry.) The result of the fault analysis is a list of the raw circuit faults caused by the defects. Below we describe the basic elements of the fault analysis procedure.

5.4.2. Fault Analysis

The fault analysis phase of Monte Carlo yield simulation identifies which of the basic types of circuit faults (i.e., short, open, open device, shorted device, new via, new device due to extra gate material and new device due to extra active material) is caused by each defect. In particular, after a defect has been placed on a layout, polygon operations are used in the neighborhood of the defect to determine whether the defect causes a circuit fault. These operations are guided by the defect model that describes what layers a defect can interact with, what layer combinations form transistors, contacts, etc.

We will briefly describe the procedures used for detecting the basic types of defects. We assume here that before fault analysis takes place there is a preprocessing step in which all equipotential regions in the layout being analyzed are uniquely labeled (called net numbers).

Shorts: A short circuit can occur when a disc representing extra conducting material defect touches two or more neighboring polygons on interacting layers that have different net numbers. An interacting layer is one that can be electrically connected to the defect material, as defined by the defect models. For example, an extra metal defect can electrically connect to metal wires, while an extra poly defect can electrically connect to other poly, and to diffusion through a buried contact. If a short circuit occurs, a list of the shorted net numbers is indicated.

There are two complications to this procedure. First, shorts caused by extra active (diffusion) or gate (poly) material must be saved even if they only touch one net. This information will later be used to connect the gate and terminals of new devices caused by extra active or gate material and then discarded. Second, some layers mask others. In an MOS process with self-aligned transistors, the gate layer masks the active layer. Where gate and active layers overlap a transistor channel occurs, rather than a wire. Therefore when considering shorts due to extra active material, the defect polygon must first have neighboring gate material subtracted from it.

Opens: An open circuit occurs when a missing material defect spans a wire or via on an interacting layer, splitting a net into several unconnected *branches*. An interacting layer is one that can be broken by the missing material. For example, a missing metal defect can break a metal wire, or a missing first-level via defect can open a first-level via. The defect models separately record whether a defect type can span wires or cause open vias. An example open wire and open via are shown in Fig. 5-16 and Fig. 5-17. If an open circuit occurs, the open procedure returns a list of branches for each broken net and the contours of the broken polygon are burst into individual polygons. These new polygons represent the starting point of each branch of the broken net.

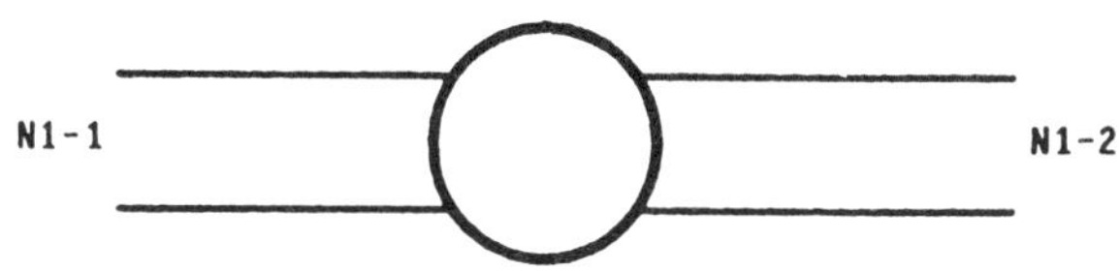

Figure 5-16: An open.

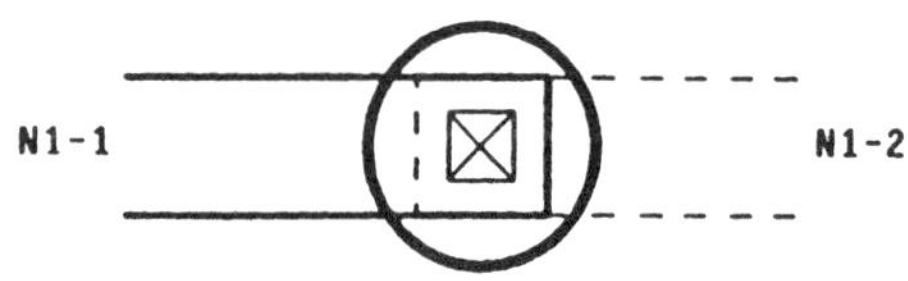

Figure 5-17: Open via.

New vias: A new via occurs when an oxide pinhole defect intersects a region where wires on conducting layers above and below the oxide layer overlap, as specified by the defect models. These overlap areas are precalculated during a preprocessing phase. New vias also occur when a junction leakage defect intersects a diffusion region. Since the new vias defects are always assumed to be points, the intersection test can be done by calculating the winding number of the point with respect to the neighboring overlap polygons. This is computationally cheaper than other intersection tests. If a new via occurs, the new via procedure returns the pair of nets shorted together.

New via faults are distinguished from shorts for two reasons. First, a new via caused by a gate oxide pinhole is not well-defined electrically. Second, the fact that oxide pinholes and junction leakage defects are always assumed to be points makes their analysis much easier.

Open devices: An open device occurs when a missing active area defect spans a transistor channel from side to side, thus breaking the channel between source and drain, as shown in Fig. 5-18. If an open device occurs, the open device procedure reports the device number. An open device is

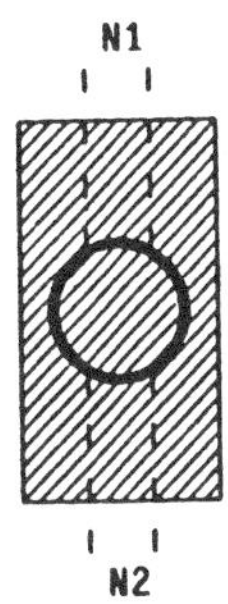

Figure 5-18: Open device.

detected by intersecting gate polygons with the defect, isolating the defect under the gate. The result may have several disjoint contours, which are burst apart into separate isolated defects, called masked defects, each of which are regarded separately. Several masked defects are created if a single defect touches more than one transistor, as shown in Fig. 5-19. If no masked defect results from the intersection, then there can be no open transistors. Channel polygons are subtracted from the masked defect. If the result has two or more contours, then the channel has been spanned by the defect, resulting in an open transistor. The subtraction order is determined by the need to avoid problems with coincident defect and polygon edges.

Shorted devices: A shorted device occurs when a missing gate (poly) defect

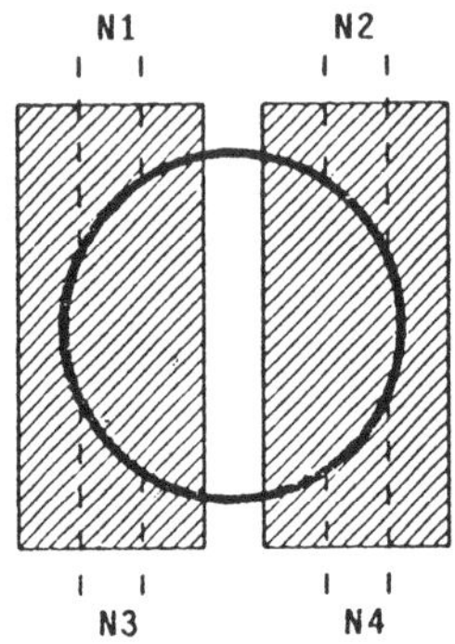

Figure 5-19: Multiple transistor defect.

spans the channel all the way from source to drain, allowing diffusion material to short between source and drain, as was shown in Fig. 5-20. If a shorted device occurs, the shorted device procedure reports the device number.

The active (diffusion) polygons provided by the circuit extractor have the channel regions subtracted from them, with the channel polygons provided on a special channel layer. In order to perform the analysis, the original active mask must be recreated by taking the union of the channel polygons and active polygons in the defect neighborhood. The defect is intersected with the active polygons, including the active area that forms the transistor channel, creating a masked defect. The channels are subtracted from the masked defect. If the result has two or more contours, then the transistor is shorted. The gate must be subtracted from the masked defect rather than

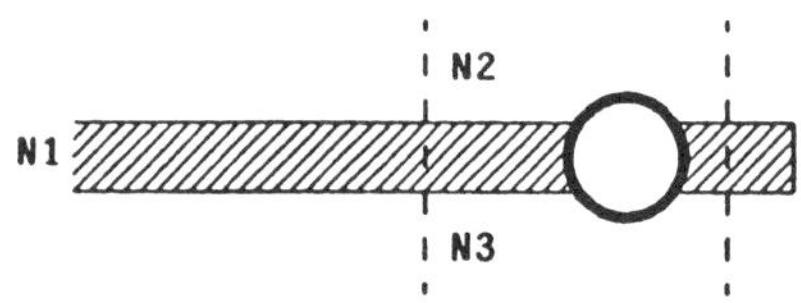

Figure 5-20: Short in the transistor channel due to missing poly.

vice versa for the following reason. If the defect covers one of the side edges of the channel, then the masked defect shares an edge with the channel. If the masked defect is subtracted from the channel, the channel will not be split in two, but only have its end removed. The defect is isolated over the active and channel because in fact a single missing gate defect can cause more than one transistor to be shorted, as shown in Fig. 5-21. When the defect is masked, the result may have several contours, corresponding to intersections with several channel/active regions. These contours are burst apart, and considered as several individual masked defects that can cause several different shorted transistor faults.

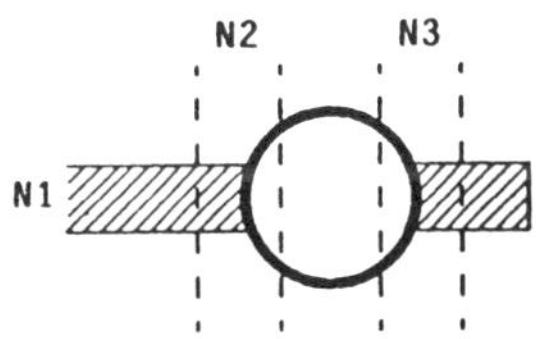

Figure 5-21: Multiple transistor short.

New gate material device: A new device due to extra gate material occurs when an extra gate (poly) defect spans an active (diffusion) wire, breaking it in two, and creating a channel connected to both source and drain, as shown in Fig. 5-22. If a new device occurs, the analysis procedure reports a new device with source and drain terminals connected to branches of a broken diffusion net. The gate terminal is connected during the fault combination phase.

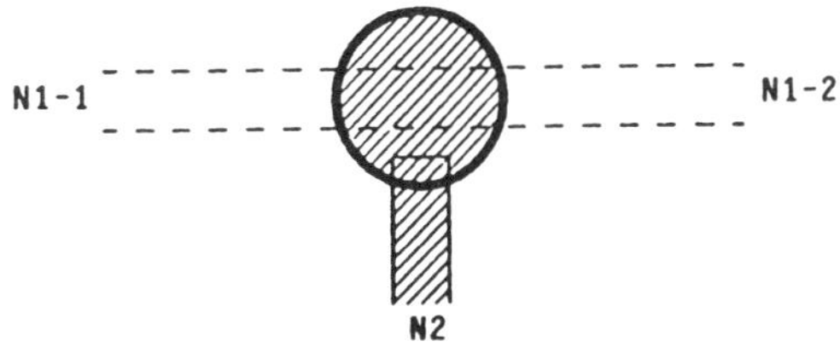

Figure 5-22: New gate material device.

A new device is created in the following cases:

- The combined channel has two terminals, and does not contain any existing channels. This is a new transistor.

- The combined channel has more than two terminals, and does not contain any existing channels. This is a new multi-terminal transistor.

- The combined channel has more than two terminals, and also touches existing channels. This is a multi-terminal transistor converted from existing transistors.

- The combined channel has less than two terminals, and does not touch existing channels. A capacitor is formed, but ignored, since it is not a DC topology change.

- The combined channel has two terminals and also touches an existing channel. If the combined channel is a different shape than the existing channel, then a device size change has occurred. If the combined channel is the same shape as the existing channel, then no size change has occurred. In either case, no DC topology change occurs, so no fault is reported.

- Otherwise no transistor is formed.

The above rules depend on the assumption that only devices with one source and one drain terminal are permitted in the original circuit.

<u>New active material device:</u> A new device due to extra active material occurs when an extra active (diffusion) defect spans a gate (poly) wire, forming a channel and a source and drain, as was shown in Fig. 5-23. If a new device occurs, the new device procedure reports a new device with terminals connected to existing nets. If the new active regions also short several nets together, then these net numbers are added to the terminal net lists in the fault combination phase.

A new active device is created in the following cases:

- The combined channel has two terminals and does not touch any existing channels. This is a new transistor.

- The combined channel has more than two terminals, and does

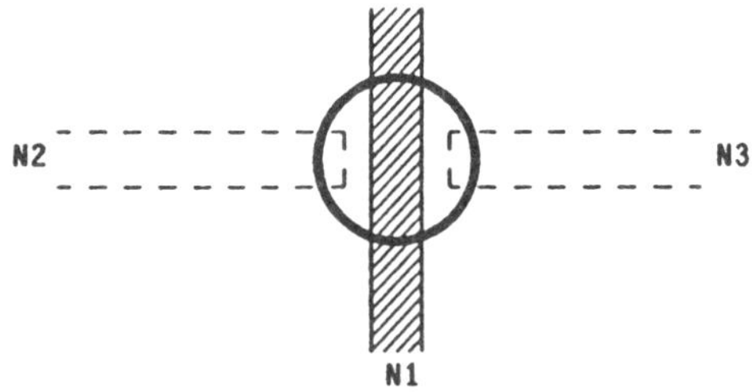

Figure 5-23: New active material device.

not touch any existing channels. This is a new multi-terminal transistor.

- The combined channel has more than two terminals and touches existing channels. This is a multi-terminal transistor converted from existing transistors.

- The combined channel has two terminals and touches existing channels. If more than one existing channel is involved, and the combined channel does not equal the original channel, then a size change has occurred. Otherwise no size changed has occurred. In either case, since no DC topology change has occurred, no fault is reported.

- Otherwise no transistor is formed.

Fault combination: Some defects may cause several different faults to occur. In some cases, these faults must be combined into a simpler fault, or discarded altogether. For example, an extra poly defect may block a metal-diffusion via. This causes an open between the metal and diffusion, and causes the metal to short to the extra poly. At the same time, the extra poly may also short to the diffusion through a buried contact. The path from the metal via through the poly to the buried contact recompletes the circuit, so no fault has occurred. Similarly, an extra poly defect can break a diffusion wire in the process of forming a new device. However the poly may connect to the diffusion on each side of the break through buried contacts, reconnecting the diffusion net.

Some faults require information from several fault analysis procedures. Information from the short procedure is used to connect up the sources and drains of a new device due to extra diffusion. Those terminals that are not connected are discarded. If the transistor does not have at least a source and drain, then it cannot change the DC topology of the network, and so is discarded. The short procedure also provides information to hook up the gate of a new device due to extra gate material.

5.4.3. VLASIC Implementation and Summarizing Discussion

The procedures described above form a core of the VLASIC (VLSI Layout Simulation for Integrated Circuit) and are implemented in approximately 20,000 lines of C code. The current version can handle, NMOS and CMOS double metal technologies[5].

The main features of VLASIC implementation are as follows:

1. VLASIC provides almost unlimited flexibility in modeling defects. All essential defect characteristics can be simulated with good accuracy.

2. VLASIC uses standard polygon packages which have many advantages and a few disadvantages. On the positive side is generality of the approach to failure analysis which allows to handle (as was shown in Sec. 5.4.2) very complete topological configurations of defects. It also allows for easy modification whenever new or changed process must be analyzed. Therefore such generality and flexibility is the cost of computations.

3. VLASIC uses the Monte Carlo approach which requires large sample sizes.

4. VLASIC can estimate probabilities of specific faults and its contribution to the yield loss.

Hence, in general VLASIC is a tool that may provide very accurate estimate of the yield loss but its application must be limited to the regular structures or functional blocks of moderate size.

---

[5] A description of the details of the implementation are out of the scope of this book and can be found in [126] and [127].

5.5. Yield Computations for VLSI Cell[6]

Both of the yield prediction methods discussed above have limitations: the virtual layout approach may be inaccurate and the Monte Carlo approach may be too expensive. In this section we present a third approach. The basic concept behind this third method is searching for elements of the IC that are sensitive to spot defects and then evaluating the probability of a fault occurring in these locations. This approach minimizes the complexity of computation while maintaining accuracy. In general, functional yield prediction can be performed on a *flat* or *hierarchical* layout. To predict functional yield on a flat layout, i.e. the layout of an entire chip, we must be able to calculate the probabilities of circuit failure and then yield simultaneously. Such a method is difficult to implement for VLSI circuits due to the complexity of circuit layout and the dependency between design rules. On the other hand, for hierarchical yield prediction, one calculates the chip yield in terms of the yields of the individual devices or cells. In this section we discuss such an approach to yield prediction. This approach employs analytical expressions rather than a Monte Carlo approach and considers yield losses due to both global and local defects, taking into account defect sizes as well as thin spatial distributions.

The method has three key steps. First, defect statistics are generated hierarchically (see Sec. 5.2). Next, the probabilities of failure (POF) due to global, and combined global and local deformations for simple layout patterns are calculated analytically. Finally, the POF's for macrocells or chips are calculated. A hierarchical scheme is used during this last step to reduce computational complexity.

5.5.1. Probability of Failure (POF) for Simple Layout Patterns

In this section, we describe methods to find probabilities of failure (POF)for simple layout patterns. We first classify the circuit failures into two types: i.e., primitive failures and joint failures. Then we define the events for these two kinds of circuit failures to occur and derive formulae to calculate their

[6]This section is based upon the paper: "Realistic yield simulation for VLSI structural failures", by I. Chen and A.J. Strojwas, IEEE Trans. on CAD, Vol. CAD-6, No. 6, pp. 965-980, 1987.

probabilities of failure. Strict analytical methods are used in this step to achieve accuracy and efficiency.

POF for Primitive Failures

Primitive failuresare the failures caused in the simplest layout patterns. These failures can be further classified into five types: shorts, breaks, lateral and vertical interlayer failures, and excessive junction leakage currents.[7]

Shorts. The source of shorts can be lithographic or nonlithographic based. As examples, a metal short is lithographic based while "*cross talk*" in poly paths and "*punch through*" in source and drain under some bias condition are nonlithographic based. In this section, we consider these two sources of shorts. Since global deformations always occur on chips, circuit shorts can be caused either by global deformations only or by the combined effect of global and local deformations. Global deformations may have more than one component for some particular mask layer. Typically, linewidth variation is the only component for all the mask layers except diffusion layer where lateral diffusion, depletion layer expansion, and bird's beak effect, also need to be taken into account (See Sec. 5.3.3). Assume that $\mathbf{g}_i$ is a random variable (with *pdf* $f_{\mathbf{g}_i}$) which represents the *i*th component of the global deformations for some particular mask layer, then the net effect of global deformations, $\mathbf{g}$ (with *pdf* $f_{\mathbf{g}}$), can be shown as:

$$\mathbf{g}=\sum_{i}^{n}\mathbf{g}_i \tag{5.37}$$

where n is equal to 4 (linewidth variation, lateral diffusion, depletion layer expansion, bird's beak effect) for the diffusion layer and is equal to 1 for the other mask layers. Since all these components are independent with one another, $\mathbf{g}$ will also be normally distributed if $\mathbf{g}_i$, $i=1$ to n, has normal distribution. More precisely, let $\mu_{\mathbf{g}_i}$ and $\sigma^2_{\mathbf{g}_i}$ denote the mean and the variance of $\mathbf{g}_i$. Then we have [86]:

---

[7]Compare terms introduced in this section with the terms defined in 5.4. They are very similar but tend to be more general.

$$\mathbf{g} \sim N\left(\sum_{i}^{n} \mu_{\mathbf{g}_1}, \sum_{i}^{n} \sigma^2_{\mathbf{g}_1}\right) \tag{5.38}$$

where $N(\mu, \sigma^2)$ denotes a normal distribution with mean μ and variance σ^2. Based upon the above fact, we can find the POF's due to global and combined deformations.

Let x_1 and x_2 be the coordinates of the layout edges, and $\mathbf{r}$ and $\mathbf{x}$ be the random variables (with *pdfs* $f_{\mathbf{r}}$ and $f_{\mathbf{x}}$) which represent the defect radius and the coordinate of defect center, respectively. If overetching is assumed to prevail, then the POF due to global deformations, i.e., P_{f0}, is given by:

$$\begin{aligned} P_{f0} &= Pr\{(x_2+\mathbf{g})-(x_1-\mathbf{g}) \le c\} = Pr\{\mathbf{g} \le \frac{(c-x_2+x_1)}{2}\} \\ &= \int_{-\infty}^{(c-x_2+x_1)/2} f_{\mathbf{g}} d\mathbf{g} \end{aligned} \tag{5.39}$$

where c represents the minimum spacing for short to occur. This equation is also valid for the case of underetching where **-g** rather than **g** is used. For the case of combined deformations, if we assume that there is already one defect located at the region between two layout patterns, then the POF (i.e. a conditional probability) can be expressed as:

$$\begin{aligned} P_{f1} &= Pr\{(\mathbf{x}-\mathbf{r}+\mathbf{g})-(x_1-\mathbf{g}) \le c \quad \textbf{and} \quad (x_2+\mathbf{g})-(\mathbf{x}+\mathbf{r}-\mathbf{g}) \le c\} \\ &= \int_{-\infty}^{\infty} f_g d\mathbf{g} \int_{(x_2-x_1+4\mathbf{g}-2c)/2}^{\infty} f_{\mathbf{r}} d\mathbf{r} \int_{x_2+2\mathbf{g}-\mathbf{r}-c}^{x_1-2\mathbf{g}+\mathbf{r}+c} f_{\mathbf{x}} d\mathbf{x} \end{aligned} \tag{5.40}$$

where **g**, **r**, and **x** are independent of each other so that joint probability density function $f_{\mathbf{grx}}$ can be decoupled into the product of $f_{\mathbf{g}}$, $f_{\mathbf{r}}$, and $f_{\mathbf{x}}$.

In general, to calculate P_{f0} and P_{f1}, typical numerical integration methods, such as Euler methods (forward or backward), trapezoidal rule, or Simpson rule [15], can be applied.

Breaks. Like shorts, breaks can be lithographic or nonlithographic based.

For example, poly breaks are lithographic based while "*metal migration*" is non lithographic based. Breaks can also be caused by global deformations or combined deformations.

P_{f0} can be expressed as below:

$$\begin{aligned} P_{f0} &= Pr\{(x_2-\mathbf{g})-(x_1+\mathbf{g}) \le c \quad \mathbf{or} \quad (y_2-\mathbf{g})-(y_1+\mathbf{g}) \le c\} \\ &= Pr\{\mathbf{g} \le (x_2-x_1-c)/2\}+Pr\{\mathbf{g} \le (y_2-y_1-c)/2\} \\ &\quad -Pr\{\mathbf{g} \le (x_2-x_1-c)/2 \ \ \mathbf{and} \ \ \mathbf{g} \le (y_2-y_1-c)/2\} \end{aligned} \tag{5.41}$$

If we assume that (x_2-x_1) is less than or equal to (y_2-y_1), then Eq. (5.41) becomes

$$P_{f0}=Pr\{\mathbf{g} \le (x_2-x_1-c)/2\} = \int_{(x_2-x_1-c)/2}^{\infty} f_{\mathbf{g}} d\mathbf{g} \tag{5.42}$$

Similarly, the POF due to combined deformations, P_{f1}, can be shown as below:

$$\begin{aligned} P_{f1} &= Pr\{\mathbf{x}-(\mathbf{r}+\mathbf{g})-(x_1+\mathbf{g}) \le c \ \ \mathbf{and} \ \ (x_2-\mathbf{g})-(\mathbf{x}+\mathbf{r}+\mathbf{g}) \le c\} \\ &= \int_{-\infty}^{\infty} f_{\mathbf{g}} d\mathbf{g} \int_{(x_2-x_1-4\mathbf{g}-2c)/2}^{\infty} f_{\mathbf{r}} d\mathbf{r} \int_{x_2-2\mathbf{g}-\mathbf{r}-c}^{\mathbf{r}+c+2\mathbf{g}+x_1} f_{\mathbf{x}} d\mathbf{x} \end{aligned} \tag{5.43}$$

Lateral Interlayer Failures. Lateral interlayer failures are failures in lateral direction between two or more different mask layers. As examples, a transistor failure due to significant misalignment between poly and diffusion is a two-layer failure; a failure between metal to contact-cut to poly so that metal and poly cannot conduct with each other is a three-layer failure. All failures involving three or more layers can be decomposed into sets of two-layer failures. All two-layer failures are assumed to be caused mainly due to linewidth variations of the two masks involved and misalignments between them. Local deformations are neglected in this case due to small probability of occurrence although combined global and local deformations still account for most of the failures in each layer.

To illustrate how to find the POF of the lateral interlayer failure, consider the example of contact-cut to poly, as shown in Fig. 5-24. If $\mathbf{g}_1$ and $\mathbf{g}_2$ represent the net global deformations in layers 1 and 2, respectively, and $\mathbf{m}_{\mathbf{x}}$

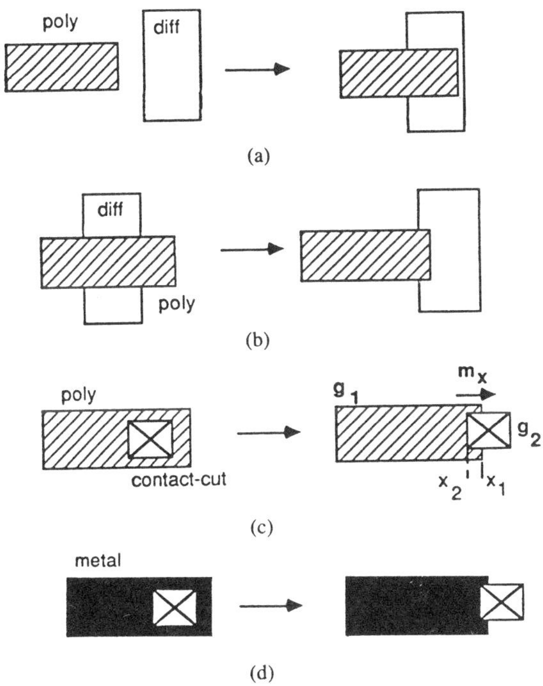

Figure 5-24: Four types of lateral interlayer failures.

represents the misalignment in X direction, then the POF can be expressed as:

$$P_f = Pr\{(x_1 - \mathbf{g_1}) - (x_2 + \mathbf{g_2} + \mathbf{m_x}) \leq c\}$$
$$= \int_{-\infty}^{x_2 - x_1 + c} f_{\mathbf{w}} d\mathbf{w} \tag{5.44}$$

where

$$\mathbf{w} = -\mathbf{g_1} - \mathbf{g_2} - \mathbf{m_x} \tag{5.45}$$

Note that in Fig. 5-24, all the four failure types are shown in +X direction. Since the misalignment may also have components in -X and ± Y directions, we also need to consider the possibility for these four failure types occurring on the other three quadrants. To do so, all the four patterns in Fig. 5-24 are first rotated clockwise by 90, 180, and 270, degrees, respectively. Then the corresponding POF's can be found quickly similarly to Eq. (5.44) by replacing suitable coordinates.

Vertical Interlayer Failures. Vertical interlayer failures are failures between two different layers in vertical direction. The most significant one is a pinhole which is a short of two vertical layers through oxide. For pinholes, the effects of global deformations need to be taken into account because the overlapping area is affected by the linewidth variations of the two masks and the misalignment. Consider the example shown in Fig. 5-25, where, within some macrocell, two layout patterns from two different mask layers are overlapping.

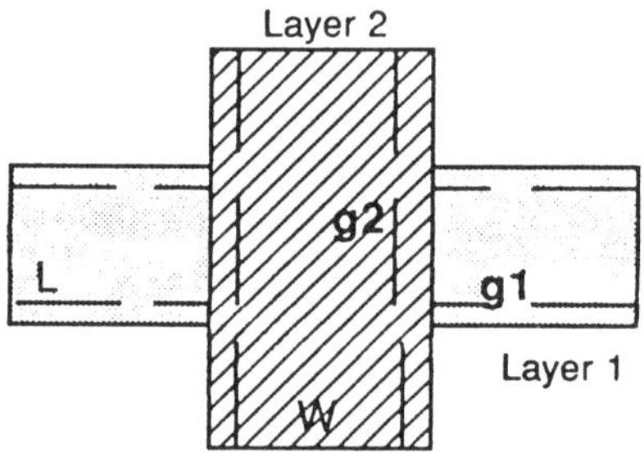

Figure 5-25: Vertical interlayer failure.

If L and W denote the length and the width of the nominal overlapping area, respectively, then the actual overlapping area A_o becomes

$$A_o = (L - 2\mathbf{g}_1)(W - 2\mathbf{g}_2) \tag{5.46}$$

where $\mathbf{g}_1$ and $\mathbf{g}_2$ denote the global deformations of mask layers 1 and 2, respectively. Since the overlapping area is very small, we can assume that the number of pinholes, if there are any, is at most one. Thus, from Eq. (5.18) and Eq. (5.19), the probability of one pinhole in A_o is

$$P_{d1} = 1 - \sum_{i=0}^{D} (1 - \frac{A_o}{A_c})^i P_{ti} \tag{5.47}$$

where D is the number of pinholes per chip and A_c is the area of the macrocell. Note that the existence of any pinholes causes a failure. As a result, the POF will then be equal to the probability of one pinhole, i.e.

$$P_{f1}=P_{d1} \tag{5.48}$$

<u>Excessive Junction Leakage Currents</u> Junction leakage currents occur in the p-n junction which is formed from the diffusion or implantation of some type of dopant into a substrate with a different type of dopant. Thus, only the diffusion mask layer is involved in this case. To find this POF, the method for vertical interlayer failures can be applied. However in this case, both $\mathbf{g}_1$ and $\mathbf{g}_2$ are replaced by $\mathbf{g}$ which is the net effect of global deformations for diffusion layer.

Joint Failures

Joint failures are the failures caused by any meaningful combinations of primitive failures. They can result from global deformations only or combined deformations as in primitive failures.

<u>Joint Failures Due to Global Deformations.</u> Primitive failures within a joint failure can be either independent or dependent on each other. As an example, if $\mathbf{F}_i$ and $\mathbf{F}_j$ denote the primitive failures of "poly short" and "metal break", respectively, then we can define a joint failure of these two primitive failures as $(\mathbf{F}_i \cap \mathbf{F}_j)$. Since the design rules for poly spacing in the events of primitive failures $\mathbf{F}_i$ and $\mathbf{F}_j$ are independent, this joint failure can be decoupled into two independent sub circuit failures, i.e. "poly short" and "metal break". The POF for this joint failure can then be obtained from the product of the POF's of these two primitive failures, i.e.

$$P_f = Pr\{\mathbf{F}_i \cap \mathbf{F}_j\} = Pr\{\mathbf{F}_i\}\ Pr\{\mathbf{F}_j\} \tag{5.49}$$

<u>Joint Failures Due to Combined Deformations.</u> For combined deformations, the components of a joint failure can also be independent or dependent. As an example, the joint failure of "poly short" and "metal short" are independent since, in addition to the independency of the design rules about the spacing in poly and metal, the defect sizes of these two mask layers are also independent. Thus, the POF of this joint failure is equal to the product of those two primitive failures.

Hence, as illustrated above, strict analytical methods rather than typical Monte Carlo-based methods can be used to find POF's for simple layout patterns. It is not difficult to derive the integration formulae for primitive failures while for joint failures, particularly due to combined deformations and dependent primitive failures, the task becomes quite involved. However, by solving this complex problem, significant gain in efficiency can be obtained. This is because the computational cost for an analytical approach is mainly due to numerical integrations and these integrations can be performed very efficiently. Thus the savings in CPU time can be quite substantial.

5.5.2. POF for Macrocells

In this subsection, we describe an algorithm for determining POF's for macrocells or chips. In general, it is very difficult to find the POF for a chip accurately and efficiently at the same time. One reason for this is that the layout of a chip is quite complex so that the number of possible failures is large. Also, many of the design rules are correlated with one another, further complicating the problem. As a result, the computational cost will be prohibitively high if the POF is calculated in a flat manner. To reduce this complexity, one can use a method which first decomposes the POF of a chip into two components, i.e. global deformations and combined deformations [18]. Next, for each component, in [18] a hierarchical model for finding the corresponding POF can be developed, see [18] for details. Under this model, a circuit in a chip is decomposed into several sub-circuits, depending on the layout complexity. Within each level, one can then perform yield simulation for all mask layers. This procedure starts from the simple layout patterns, as stated in the last paragraph, and then up to the cell and chip level. Finally, the POF of a chip can be obtained from those two components.

POF due to Global Deformations

POF's due to global deformations may not be so significant as compared to combined deformations. However, they will become more and more important in the future due to more tight particle control in clean rooms and more stringent design rules. The failure types relevant to this kind of POF can be shorts, breaks, and lateral interlayer failures. Since global deformations are assumed to be constant within a chip, failures due to these deformations will always occur on the layouts with nominal minimum spacings, widths, and interlayer spacings for shorts, breaks, and interlayer failures, respectively. To prove this, consider the example of a short, as shown in Fig. 5-26, where layouts with two different spacings are highlighted.

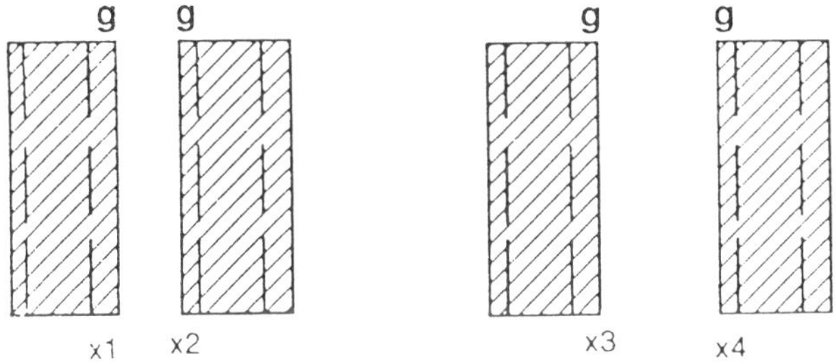

Figure 5-26: Short due to minimum layout spacing.

Then the event of failure for these two sets of patterns is equal to the union of these two individual events of short. This can be expressed as:

$$\begin{aligned} POF &= Pr\{(x_2+\mathbf{g})-(x_1-\mathbf{g}) \leq c \ \ \mathbf{or} \ \ (x_4+\mathbf{g})-(x_3-\mathbf{g}) \leq c\} \\ &= Pr\{\mathbf{g} \leq (c-(x_2-x_1))/2\} + Pr\{\mathbf{g} \leq (c-(x_4-x_3))/2\} \\ &\quad - Pr\{\mathbf{g} \leq (c-(x_4-x_3))/2\} \qquad (5.50) \\ &= Pr\{\mathbf{g} \leq (c-(x_2-x_1))/2\} \qquad (5.51) \end{aligned}$$

where we assume that (x_2-x_1) is less than or equal to (x_4-x_3). This implies that to find the POF due to short, only the part of layouts with the smallest spacing needs to be considered. This fact can be extended to any N sets of patterns. Similarly, it can be shown that only the sets of layouts with minimum width and minimum interlayer spacing are required for breaks and interlayer failures.

From the above, we conclude that the problem of finding the POF due to global deformations for large layout can be significantly simplified by applying Eq. (5.39), Eq. (5.42), and Eq. (5.44) to these nominal minimum values. Based upon this fact, an algorithm for finding POF caused by global deformations was proposed [18]. Below are described the main steps of this algorithm.

Step 1: *Partition the chip into sets of macrocells.*

In this step, to keep the hierarchical model of defect statistics valid, some thresholds of the upper-bound and the lower-bound of the cell area are used in the partition.

Step 2: *Find the nominal minimum spacing, nominal minimum width, and nominal minimum interlayer spacing, of the cell for each mask layer from the IC layout.*

To achieve this step, the description of the IC layout is first transferred into CIF (CalTech Intermediate Format). Then, a special purpose circuit extractor is used to extract all the required information by analyzing the CIF description.

Step 3: *For each mask layer in the cell layout, find the total areas sensitive to shorts and breaks, respectively.*

The sensitive area is defined as a region such that if the center of a defect is located in it, a short or a break may occur.[8] This sensitive area is a function of defect size and layout geometries considered. In this section, we determine the sensitive areas based upon some threshold of defect size. Given a confidence level α, this threshold can be obtained from the defect size distribution under which $100(1-\alpha)\%$ of the defects may occur. Note that we do not need to find the total areas of interlayer spacings since we assume that the interlayer failures can only occur due to global deformations. This step is done by using a scan-line algorithm to analyze the IC layout.

Step 4: *Find the probabilities for having no defects on the total sensitive areas to short and break, respectively, for each mask layer.*

Let $A_{t,s}$ and $A_{t,b}$ denote the total areas sensitive to short and break, respectively. Then, from Eq. (5.18), the probabilities for having no defects in these two areas become

[8] Notice that "sensitive area" is equivalent to the "critical area" discussed in Sec. 5.3.

$$P_{d0,s}=\sum_{i=0}^{D}(1-\frac{A_{t,s}}{A_c})^i P_{ti} \tag{5.52}$$

$$P_{d0,b}=\sum_{i=0}^{D}(1-\frac{A_{t,b}}{A_c})^i P_{ti} \tag{5.53}$$

where A_c is the cell area and D is the number of defect in chip.

<u>**Step 5:**</u> *At the cell level, find the POF's due to short, break, interlayer failures, and joint failures, respectively, for each mask layer.*

For the cases of shorts and breaks, Eq. (5.39) and Eq. (5.42) can be applied to the nominal minimum spacing and the nominal minimum width, respectively. For the case of interlayer failures , only the meaningful interlayer combinations need to be taken into account. As examples, the interlayer failures for metal layers to contact-cut (or via) are possible while the failures for metal to poly may be meaningless. Eq. (5.44) with nominal minimum interlayer spacing can be used for the POF of those meaningful combinations. Similar to joint failures, only the meaningful combinations of any two primitive failures needs to be calculated. Joint failures with combinations of more than two primitive failures have been proven to be very small and can be neglected.

<u>**Step 6:**</u> *Find the POF of the cell.*

Let $\mathbf{F_i}$ denote the event of any possible failure (short, break, or interlayer failure) for any possible mask layer (diffusion, poly, etc) and $\mathbf{DO_i}$ denote the event for no defects in $\mathbf{F_i}$. Then the POF due to global deformations for the cell becomes

$$\begin{aligned} POF_g &= Pr\{(\mathbf{F_1} \cap \mathbf{DO_1}) \cup \ldots \cup (\mathbf{F_i} \cap \mathbf{DO_i}) \ldots\} \\ &= \sum_i Pr\{\mathbf{F_i} \cap \mathbf{DO_1}\} - \sum_{i,j,i \neq j} Pr\{(\mathbf{F_i} \cap \mathbf{DO_i}) \cap (\mathbf{F_j} \cap \mathbf{DO_j})\} + \ldots \\ &= \sum_i (P_{d0})_i (P_{f0})_j - \sum_{i,j,i \neq j} (P_{d0})_i (P_{d0})_j (P_{fj0})_{ij} + \ldots \end{aligned} \tag{5.54}$$

where P_{d0} denotes the probability of no defects, P_{f0} denotes the POF of primitive failures, and P_{fj0} denotes the POF of joint failures. All these three terms can be obtained from Steps 4 and 5, respectively. Note that, a threshold can be used to bypass the insignificant combinational terms to save computational cost. In this case, only the expansion terms with one or two primitive failures are considered.

POF due to Combined Deformations

Recall that we have assumed that shorts, breaks, and vertical interlayer failures may be caused by combined effects. The POF of the vertical interlayer failures can be obtained easily by integrating the density of pinholes over all possible overlapping areas. Thus, we will focus on the POF's due to shorts and breaks in this section.

In general, the problem of finding POF's due to combined deformations is very complicated. This is because, in addition to the design rules, the sizes of the spot defects are distributed according to the same *pdf* and hence are correlated. As a result, the computational cost will become prohibitively high due to huge number of integrations required, if we consider the problem in a flat manner.

One way to reduce the problem of complexity is to decrease the number of patterns to be analyzed. Toward this end, we can approximate some number of layout patterns for the same mask layer by a simplified pattern. Then, the POF due to the combined deformations for these layout patterns can be approximated by just calculating the POF due to one defect on this simplified pattern. We call this simplified layout pattern to be an *equivalent layout*. Note that equivalent layout is in concept similar to the virtual layout [66, 67] described in section 5.3.1. However, while the virtual layout is applied to the entire chip, the equivalent layout is more microscopic. It is built in a bottom-up fashion with the precise thresholding to save computational cost while still maintaining the desired accuracy.

Note now that since layout patterns can be simplified by applying the equivalent layout, we can combine this concept with the hierarchical model to find the POF in the cell and chip levels. Below we describe the main steps of the algorithm.

Step 1: *Partition the macrocells into sets of intermediate logic gates*

Some cells have areas similar to that of intermediate logic gates. In this case, partitions are not required and the cell can be treated as an intermediate logic gate. Thus, some threshold for area is used to determine whether partitions are needed.

Step 2: *Find the equivalent layouts for shorts and breaks, respectively, in each mask layer of each logic gate*

For circuit shorts, threshold for spacing between two layout patterns is used while for breaks, threshold will be the width of layout patterns.

Step 3: *Find the probability of 1 defect in the equivalent layout of each mask layer within each partitioned logic gate*

It can be shown that the number of defects occurring in the equivalent layout can be assumed to be at most equal to 1, if the area of the equivalent layout, A_{eq} $(=L_{eq} W_{eq})$, satisfies the following criterion:

$$\frac{A_c}{A_{eq}} \geq D \tag{5.55}$$

where A_c is the cell area and D is the number of defects per chip. This is because the number of the defects on the cell, A_c, can be at most equal to D due to clustering (although the probability may be very small). Thus, if the ratio A_c/A_{eq} is smaller than or equal to D, then the average number of defects on A_{eq} may be greater than 1, due to the assumption of uniform distribution of defects within a cell. As a result, if A_{eq} satisfies the above criterion, Eq. (5.19) can be applied to calculate the probability of 1 defect in the equivalent layout:

$$P_{d1} = 1 - \sum_{i=0}^{D} (1-\frac{A_{eq}}{A_c})^i P_{ti} \tag{5.56}$$

Step 4: *Find POF's due to shorts and breaks, respectively, in each equivalent layout for each mask layer*

This can be done by applying Eq. (5.40) and Eq. (5.43) to the corresponding equivalent area, respectively.

Step 5: *Find POF's due to shorts and breaks, respectively, in each logic gate for each mask layer*

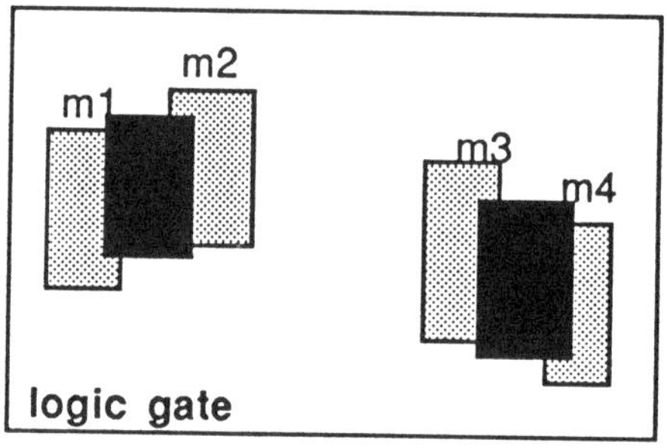

Figure 5-27: A partitioned logic gate.

Consider the case of short, as shown in Fig. 5-27. If A_{eq} denotes the equivalent layout for short and $(\mathbf{m_a}, \mathbf{m_b})$ represents the layout patterns with the spacing A_{eq}, then we can define $\mathbf{D1_{eq}}$ and $\mathbf{F_{eq}}$ to be the events of "1 defect in A_{eq}" and "short in $(\mathbf{m_a}, \mathbf{m_b})$", respectively. Then the POF due to combined deformations for this equivalent layout becomes:

$$P_{f1} = Pr\{\mathbf{D1_{eq}} \cap \mathbf{F_{eq}}\} = P_{d1,eq} \, P_{f1,eq} \tag{5.57}$$

where $(\mathbf{m_a}, \mathbf{m_b})$ is the layout patterns with the spacing A_{eq}, and $P_{d1,eq}$ and $P_{f1,eq}$ can be obtained from Steps 3 and 4, respectively. This method can be also applied to break for all the mask layers. Note that, joint failures, in this case, can be neglected due to the very small value of the product of the probabilities for 1 defect in both individual failure events.

Step 6: *Find the POF of the macrocell*

Consider Fig. 5-28, where a macrocell with partitioned logic gates is highlighted. Let $\mathbf{F_{ij}}$ denotes the event of failure on the equivalent layout of the ith mask layer in the jth logic gate and $\mathbf{D1_{ij}}$ represents the event of 1 defect on this equivalent layout. Then the POF for a macrocell becomes:

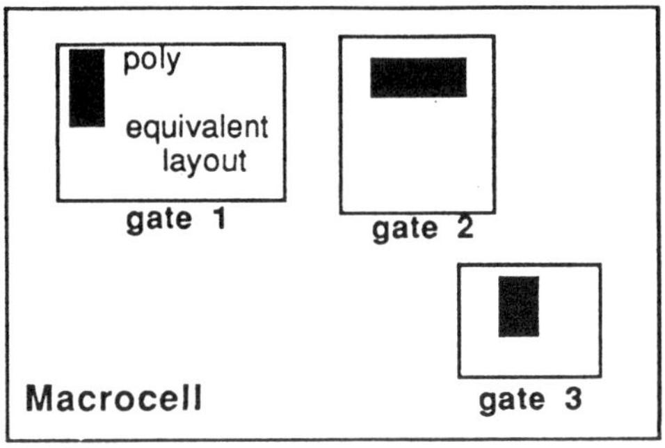

Figure 5-28: A functional cell with partitioned gates.

$$POF_c = Pr\{(\mathbf{F_{11}} \cap \mathbf{D1_{11}}) \cup \dots \cup (\mathbf{F_{ij}} \cap \mathbf{D1_{ij}}) \dots\}$$

$$= \sum_{i,j} Pr\{\mathbf{F_{ij}} \cap \mathbf{D1_{ij}}\} - \sum_{i,j,k,l} Pr\{(\mathbf{F_{ij}} \cap \mathbf{D1_{ij}}) \cap (\mathbf{F_{kl}} \cap \mathbf{D1_{kl}})\} + \dots$$

$$= \sum_{i,j} (P_{d1})_{ij} Pr\{\mathbf{F_{ij}}\} - \sum_{i,j,k,l} (P_{d1})_{ij} (P_{d1})_{kl} Pr\{\mathbf{F_{ij}} \cap \mathbf{F_{kl}}\} + \dots \qquad (5.58)$$

where P_{d1} and $Pr\{\mathbf{F_{ij}}\}$ are already obtained form Steps 3 and 4, respectively. To find the second term in the RHS of Eq. (5.58), a large number of the possible joint failures need to be calculated. Fortunately, the computational cost is not prohibitively high since, in most cases, the two primitive failures of a joint failure are independent with each other. More precisely, we have

1. If $\mathbf{i} = \mathbf{k}$ and $\mathbf{j} = \mathbf{l}$, then $Pr\{\mathbf{F_{ij}} \cap \mathbf{F_{kl}}\}$ is equal to 0 since $\mathbf{F_{ij}}$ and $\mathbf{F_{kl}}$ are the same event.

2. If $\mathbf{i} \neq \mathbf{k}$, then $\mathbf{F_{ij}}$ and $\mathbf{F_{kl}}$ are independent and $Pr\{\mathbf{F_{ij}} \cap \mathbf{F_{kl}}\} = Pr\{\mathbf{F_{ij}}\} Pr\{\mathbf{F_{kl}}\}$, where $Pr\{\mathbf{F_{ij}}\}$ and $Pr\{\mathbf{F_{kl}}\}$ are already found in Step 4.

3. Otherwise, $\mathbf{F_{ij}}$ and $\mathbf{F_{kl}}$ are dependent and appropriate formulae can be used to find $Pr\{\mathbf{F_{ij}} \cap \mathbf{F_{kl}}\}$.

Final POF for a Macrocell

At this stage, we already know POF_g and POF_c of a macrocell. Note that POF_g and POF_c are defined in such a way that both of them are independent. This is because conditional probabilities for no defect and for defects to occur in some regions of interest are already taken into account for POF_g and POF_c, respectively. Besides, from the investigation of these two components, all the possible failures which are not negligible are included in our methodology. Thus, the final POF for a macrocell is given by

$$POF = POF_g + POF_c. \tag{5.59}$$

5.5.3. Implementation

A prototype software system called **RYE** [18] was written in C language. Currently, it can find probabilities of failure for macrocells for both NMOS and CMOS technologies. It considers the specific IC layout and accounts for most fault mechanisms caused by global geometrical variations and local defects.

Two key aspects should be emphasized. First, a model for defect spatial distributions in a hierarchical manner has been derived. This model, unlike the methods which use curve-fitting based statistics, is microscopic and analytically-based and includes a new approach to account for local clustering at cell level. Second, a hierarchical method for POF calculation has been developed. This method allows us to calculate the POF's of primitive and joint failures for simple layout patterns. Then, based upon these POF's and defect spatial distributions, probabilities of failure at more complex circuit levels, such as logic gates and macrocells, can be determined in a bottom-up fashion. In [18] a number of examples are shown which illustrate efficiency of the yield computation method implemented in RYE. The efficiency was high indicating that functional yield loss can be effectively computed for relatively large ICs.

Chapter 6
Computer-Aided Manufacturing[1]

In this chapter we present a methodology for statistical process control of VLSI fabrication processes. We formally introduce a general framework for a Computer Aided Manufacturing system that can be used to monitor, diagnose and control IC manufacturing. We formulate the task of process control as one of profit maximization and develop the associated objective function and the constraints for a number of manufacturing scenarios.

6.1. Motivation

Any VLSI manufacturing line is subjected to two types of constraints: external and internal. External constraints are specified by the customer and include bounds on the output performances of the IC's and delivery dates. Internal constraints, on the other hand, consist of the restrictions imposed by the manufacturing process and the technology. These include processing times for the different fabrication steps, resource capacities, mean failure rates, as well as inherent random process fluctuations associated with each step. Together these constraints affect the number of *acceptable* chips that can be delivered to the customer.

The aim of any VLSI manufacturer is to maximize the total profit while meeting all the constraints listed above. Profit is affected by a number of factors. The inherent process fluctuations cause significant yield losses and that in turns results in a drop in the total profit of the IC fabrication line.

---

[1]This chapter is based on the papers: "Statistical Control of VLSI Fabrication Processes: A Framework" by P. K. Mozumder, C. R. Shyamsundar and A. J. Strojwas, and "Statistical Control of VLSI Fabrication Processes: An Implementation", IEEE Trans. on Semiconductor Manufacturing, vol. 1, May, 1988.

Maintaining a high throughput rate alone may not be sufficient to offset the impact of a low yield. Fabrication process times cannot be assumed to be constant, especially since some of the fabrication steps may have to be redone complicating the job of scheduling individual process steps. Fabrication costs are affected by rework which affects the total profit. In addition, IC demand fluctuations and scheduled and unscheduled downtimes of the resources available to a manufacturing line make it difficult to arrive at a policy to maximize the total profit.

The total profit is affected by the different fabrication/assembly costs and revenue associated with the fabrication process [98, 102]. These costs and profits are in turn dependent on several aspects of the particular fabrication process that a lot undergoes. Some of the processing steps are performed on entire lots and hence the total cost associated with such steps will not depend significantly on the number of wafers. Other processing steps are performed on individual wafers and the costs associated with such steps depend on the number of wafers. The above costs are referred to as *front end costs* . After the wafer processing stage, wafers are divided into individual chips each of which are probe-tested. The chips that pass these tests are assembled, packaged into IC's and then undergo functional tests. The costs associated with these testing and assembly operations are grouped under the category of *back end costs*. Note that the probe and functional tests are performed on individual chips as opposed to the fabrication operations which are performed either on individual wafers or on lots of wafers. In a fabrication process with a very high yield, the back end costs, which are approximately proportional to the number of good chips, dominate. This is because, on a per chip basis, the costs associated with chip level operations are much higher than those associated with wafer/lot level operations. However, if the yield were very low (which is the case for a complicated VLSI manufacturing process), the front end cost per good chip may be comparable to the corresponding back end cost. Clearly, an IC manufacturer needs to know the profit per good chip in order to maximize the total profit from the IC manufacturing line.

A number of decisions concerning the continued processing of the lot undergoing fabrication are made during the course of the fabrication process. These include rejection of the lot, acceptance of the lot for further processing and rework of the previous step. The total profit serves as the objective to be optimized while making these decisions. For instance, a lot is rejected if the total profit expected from the lot does not justify the fabrication cost that will be incurred if the lot were to be processed completely. The fabrication process has to be controlled effectively in order to realize the objective (maximizing the total profit). Control procedures can be *on-line* or *off-line*. While on-line control deals with *process monitoring* and *quality control* , off-line control deals with *quantity control* for throughput maximization, *diagnosis* which is the method of identifying the fluctuations which cause a yield drop, and *redesign*.

The total profit depends on a number of parameters, the fabrication yield being one of the salient among these. Other parameters of interest include process stability and robustness which affect the lot to lot variation of the yield. In complex VLSI processes, the total production yield and hence the profit may be significantly affected by the variability of a single critical circuit performance. These mechanisms for profit loss can be countered by designing the process as well as the circuit so that the critical parameters, on which the profit depends, are least sensitive to process fluctuations. The design can then be supplemented by appropriate diagnosis and process control. Thus, the lot being processed is monitored via in-line and test structure measurements and then suitably controlled based on the observed deviations of the process conditions from normal.

Present day VLSI manufacturing lines rely heavily on several forms of computer aided manufacturing (CAM) systems to perform some of the off-line and on-line control procedures discussed above. Specifically, these CAM systems perform a number of *passive* functions such as implementing processing schedules, acquiring large samples of in-process data and postprocessing the data without making full statistical inferences about the process. In this chapter we describe an improved CAM system, called CMU-CAM system, that is capable of performing not only the passive functions presently in use but also a number of *active* functions such as process diagnosis and statistical control based on the gathered data. This system attempts to integrate the independent phases of IC fabrication, namely design (process and layout), wafer processing, testing, diagnosis and control, into a set of strongly interacting subsystems for the purpose of maximizing the profit from the fabrication line (Fig. 6-1).

6.2. Overview of the CMU-CAM System

The CMU-CAM system (Fig. 6-1) is composed of three strongly coupled blocks: design for manufacturability, statistical process control and diagnosis. The objective of design for manufacturability is to develop a

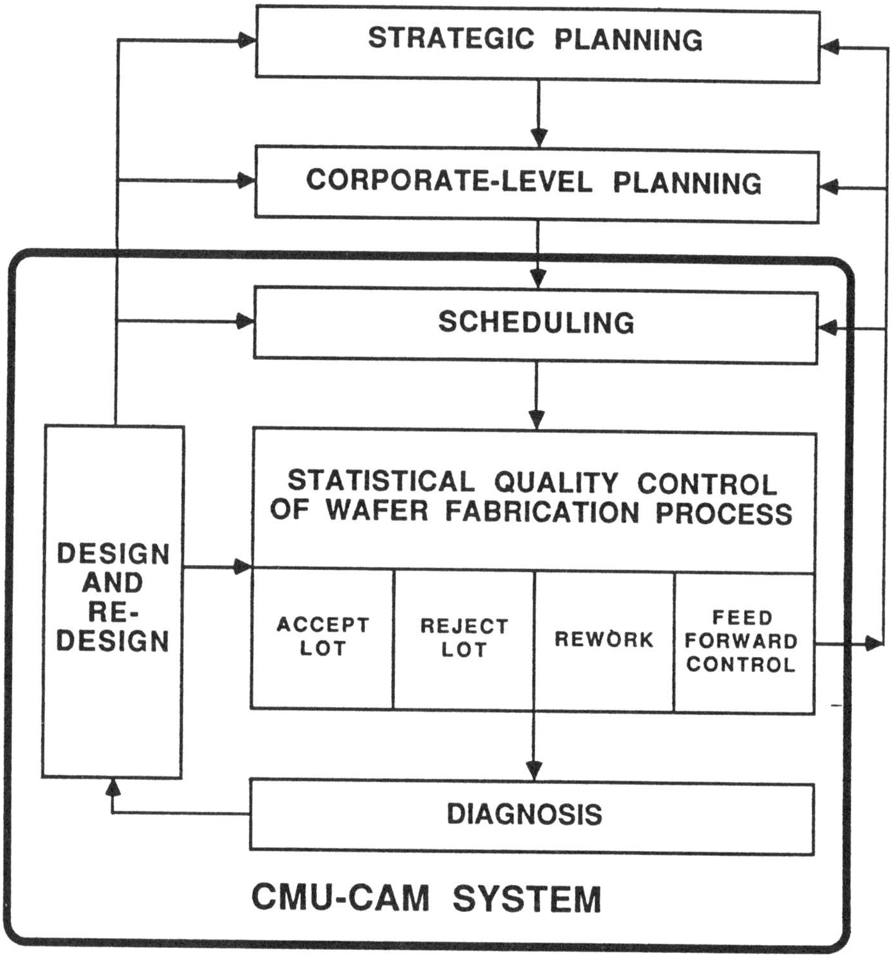

Figure 6-1: CMU-CAM system capabilities.

design that minimizes the sensitivities of the output performances of the fabricated ICs to the random disturbances affecting the manufacturing process. Several methods of performing this optimization exist in literature, two of which have been used in the context of statistical process control: design centering and quality improvement techniques using *signal to noise ratios* [119, 11]. Design centering concepts from [23, 113, 12] have been implemented in the CMU-CAM system for process control purposes [77, 102] and can also be used for IC design optimization. Data gathered by the CAM system together with statistical process and circuit simulators enable the characterization of the fabrication line which plays a crucial role in design and synthesis of the IC. The design phase may also involve the manufacturing of test lots, the data from which can be used to fine tune the design. The data gathered during the design phase is later used to schedule lots and monitor them.

Design is followed by wafer fabrication which consists of a large number of complex processing steps. During the fabrication process, a number of in-line measurements (such as oxide thickness, sheet resistances and junction depths) are made after each salient processing step. The data from these measurements enables the system to determine whether the lot under fabrication is subjected to the expected process conditions and disturbances. Based on this knowledge the process can be diagnosed to determine the causes of the deviations from the designed *process trajectory* . The process trajectory is the state of the process determined from a set of in-line measurements that allows the system to evaluate the outcome of the lot under observation. (A more formal definition of the process trajectory is presented in Section 5.) The statistical process control subsystem can then be invoked to use the data gathered during the fabrication steps to bring the process back to its designed trajectory. Process and equipment diagnosis play an important role in correcting and updating process designs from lot to lot. This is especially useful during initial stages of manufacturing an IC, i.e., when the process has not completely stabilized. We now turn to a discussion of statistical process control in the framework of the proposed CAM system.

6.3. Statistical Process Control: The Unified Framework

Statistical process control entails monitoring and then controlling the fabrication process to optimize a desired objective. In any fabrication line, two different objectives are used and correspondingly two forms of control are possible, viz., quantity control and quality control. Quantity control systems maximize throughput, i.e., maximize the number of wafers fabricated in a given period of time, whereas quality control systems maximize individual lot yields. Scheduling the lots in a fabrication process is very important since different process flows may be required for different IC's, and at any time a number of processes compete for available resources. The quantity control approach works well for a high volume manufacturing process with a high and stable yield, but is not entirely suitable for a VLSI fabrication process where the yield is typically much less than 100% and not constant (see Fig. 6-2).

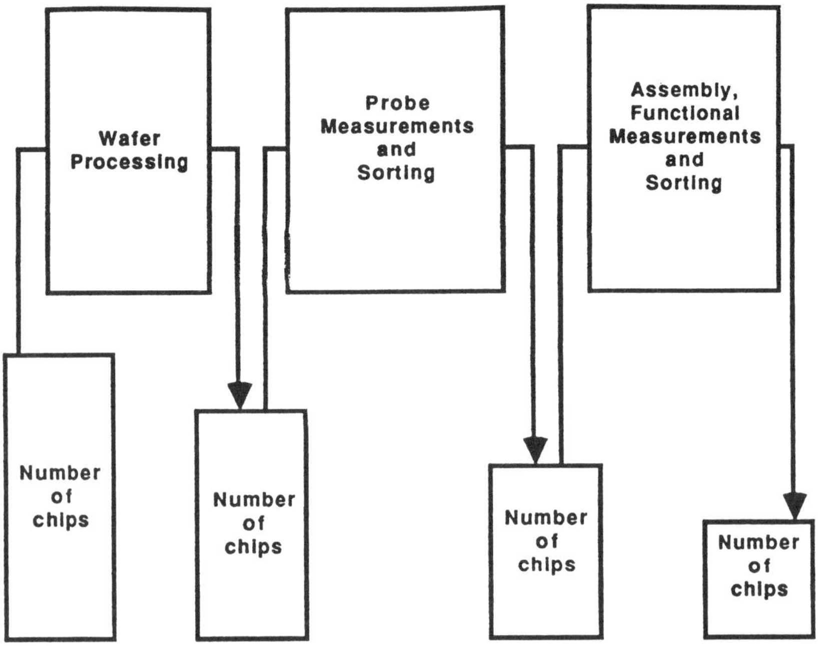

Figure 6-2: Yield loss in an IC fabrication process.

The CMU-CAM system addresses profit maximization within a unified framework in which both quantity and quality controls are considered. Depending on the predicted value of the profit, four possible decisions can be made to minimize the total fabrication costs thus maximizing profit:

1. To let further processing of the given lot continue without any modification to the process trajectory (the normal process), when the predicted fabrication yield is above its *threshold of acceptability*.

2. To explore the possibility of corrective measures, by manipulating the future process parameters, whereby the expected profit is set above the threshold of acceptability. This is done when the predicted profit fails to satisfy the previous criterion. If a feasible solution exists, the lot is processed further

with the new set of process parameters. The process of determining and manipulating the future process parameters is *feed forward control* .

3. To *rework* a previous faulty process step. Rework is not always possible for all process steps (e.g., implantation or thermal redistribution) especially where the step cannot be immediately observed via in-lines.

4. To reject the lot from further processing, when the expected profit from the lot does not justify its further processing. The lot is rejected in the case where it is not cost effective to continue further processing or take corrective measures because of large process deviations from the designed (or normal) trajectory.

Process yield maximization, as attempted by the quality control system, does not imply a profit maximization. A lot which is rejected at an intermediate stage results in zero yield, but saves the cost of future processing steps as well as frees up the fabrication equipment that could then be used to produce a lot with a much higher yield. So while no income can be derived from the rejected lot, the cost of manufacture is minimized. It is also possible for the system, based on the feed forward control, to determine a new set of values of controllable process parameters in subsequent steps that would result in a higher yield that would make further processing of the lot profitable. Therefore, the objective of the control system will be profit maximization, as opposed to yield or throughput maximization, when trying to schedule, control and optimize the fabrication process.

At the start of the fabrication process, lots are scheduled using production planning algorithms to maximize the profit. A static distribution of circuit performances and hence yield is used by the planning algorithm to maximize the total profit. The values of the moments of the circuit distribution are based on prior experience, test runs, as well as the information gathered during the design process. However, process disturbances can differ significantly from the values estimated by the process characterization algorithms, as well as vary significantly from lot to lot. Hence it may become necessary to modify the plan after a particularly good or bad lot has been processed. It is therefore imperative to monitor each lot undergoing fabrication and make quality control decisions regarding the future of the lot. These quality control decisions may lead to modification of the processing schedule. In cases of lot rejections, especially, it is necessary for the future lots to be rescheduled so as to maximize the profit. It is evident that there should be a transfer of information between the quantity and quality control subsystems. This is carried out by a short-term scheduler. For the quantity control system the short term scheduler will determine whether a rescheduling is necessary as a result of quality control, and if so perform it. For the quality control system, it will inform the system of a change in the expected profit due to the quantity control decisions, based on the success of previous lots, the time remaining to meet the demand and present state of the lot being monitored. This information will help the quality control system to determine the thresholds of acceptability and rejection. Therefore the quality and quantity control operations can be looked upon as the inner and outer loops in the overall profit maximization problem respectively.

It is not sufficient to be able to make quality control decisions in the fabrication process when the lot to lot variations in the process are large, i.e the process is not sufficiently stable. Process diagnosis, which is the method of determining the process variations from in-lines as well as output performances, becomes an important part of the CAM system enabling redesign of the process. It can be used to update the schedules determined by the quantity control system by modifying the yield estimate used by the scheduler.

6.3.1. Profit Function

In a typical IC fabrication line, a number of different fabrication recipes being used in order to manufacture different kinds of IC's. From the viewpoint of wafer fabrication all of the IC's that are manufactured using the same recipe (fabrication, sort, assembly and test) can be considered to belong to a single product group. The front-end fabrication costs will be the same for all IC's in the same product group.

In our formulation, we have assumed that for each IC, lower and upper bounds on the demand are specified. The aim is to fabricate at least as many lots of each IC as are required to meet the lower bounds on their demands within the specified due dates. For each IC, as many lots as are required to meet the upper bounds on its demand will be scheduled. Penalties will be assessed when lower bounds are not met within the specified dates. Fig. 6-3 shows a typical demand function with five demand dates along with the lower and upper bounds on the demand at each of these dates, e.g., N^{dl}_{i1} and N^{du}_{i1} are the lower and upper bounds on the demand (for IC type i) at the first demand date, T^{dem}_{i1}.

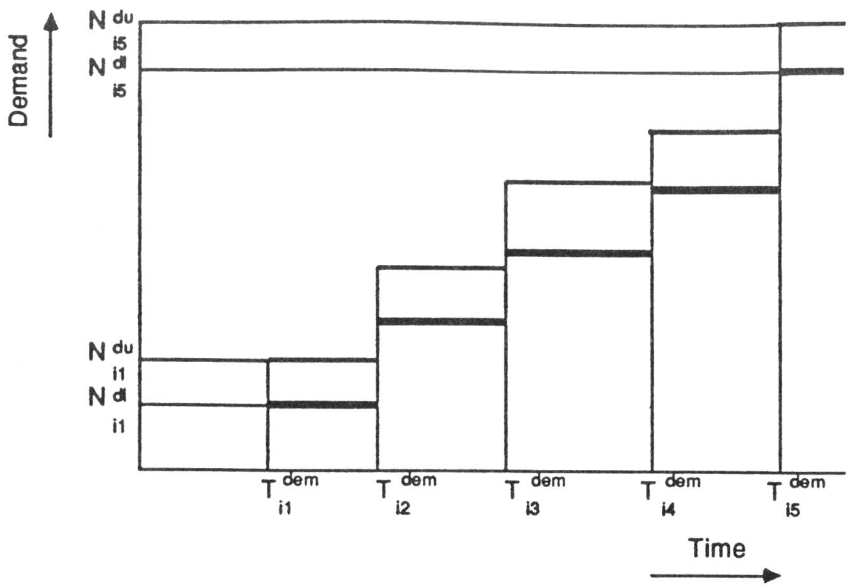

Figure 6-3: IC demand curve.

The objective function is given as the summation of the cost, penalty and revenue associated with each type of IC. Specifically, for each IC type, there is a back end cost, front end cost, revenue from the good IC's and penalty for not meeting the lower bounds on the demand at each of the demand dates.

Let T^{sch}_{im} represent the time at which the m*th* demand of the i*th* type of IC is met. Let N^{c}_{i} represent the average number of good IC's expected from each lot (of IC type i) that is scheduled. T^{sch}_{im} must not exceed the demand date T^{dem}_{im} in order to avoid the penalty for not meeting the demand. The objective function can be expressed as follows :

$$\sum_{i=1}^{N^{ic}} f_i^u - N_i^l(C_i^{fe} + C_i^{be}) - f_i^l$$

The C_i^{fe} and C_i^{be} terms refer to the front and back end costs. The penalty function f_i^l is given by

$$f_i^l = \sum_{\forall m} \mathbf{MAX}(K_{im}(T_{im}^{sch} - T_{im}^{dem}), 0)$$

There is a penalty associated with each demand that is not met by the specified date and the total penalty is the sum of the individual penalties. K_{im} is the penalty factor associated with the m*th* date for the i*th* type of IC. The revenue from the good chips is given by

$$f_i^u = C_i^{ic} N_i^l N_i^c$$

where, C_i^{ic} is the market price of the IC and C_i^c is the average number of good IC's accepted from each of the N_i^l lots that are scheduled.

6.4. CMU-CAM Software System

In this section we present the algorithms used to implement our system along with the description of their software implementations. The CMU-CAM system performs three major operations - modeling, quality control and feed forward control. In order to increase the efficiency of modeling and control, we first decompose the problem using statistical factorization techniques.

6.4.1. Decomposition

The first part of the CMU-CAD system deals with problem decomposition, where the in-lines (such as oxide thicknesses or junction depths) are clustered and then statistically factorized to identify the factor in-lines [78]. The decomposition subsystem performs two salient activities: problem decomposition, and determination of process observability and controllability. In an actual IC manufacturing line, fabrication process data gathered from test lots can be used to cluster and factorize the in-lines. We use FABRICS [81] (tuned to the desired process), as a virtual fabrication line, in conjunction with the circuit simulator SPICE.

The first step of problem decomposition consists of clustering the in-lines such that no two in-lines from two different clusters have a coefficient of correlation greater than a certain threshold value ρ_T. The clustering can be expressed as

$$\rho_{ij} < \rho_T \qquad \forall U_i \in G_p,\ U_j \in G_q \ \ (p \neq q) \tag{6.1}$$

where

G_p, G_q

are the p*th* and the q*th* clusters and

ρ_{ij}

is the multivariate correlation coefficient between U_i and U_j.

Since ρ_T is the upper bound on the value of correlation of any two elements from two different clusters, it is called the *inter-cluster correlation*. Ideally, if this value is set to zero the clusters will be independent of one another thus ensuring that there are no common process parameters or disturbances which affect more than a single cluster. However, since there exists small correlations between otherwise independent in-lines - caused by weak dependencies between certain process parameters, disturbances and in-lines - ρ_T is kept above a critical value. This ensures that the clustering algorithm is insensitive to these weak dependencies which do not significantly affect the accuracy of control and allows the system to decompose the problem into subproblems of smaller dimensionality. The exact value of ρ_T is dependent on the percentage of variance of the in-lines that needs to be accounted for, the confidence interval of the correlation coefficient arising from a finite sample size, and the error that can be tolerated due to misclassification. These factors jointly determine sample size and the lower bound on the value of ρ_T.

Principal component decomposition is applied so that n in-lines in the cluster can be represented as a linear combination of n principal factors [42]. The decomposition can be expressed as

$$U_i = \lambda_{i1} F_1 + \cdots + \lambda_{ij} F_j + \cdots + \lambda_{in} F_n \tag{6.2}$$
$$i = 1,2, \ldots, n;$$

where U_i is the *ith* in-line in the cluster to be factorized, F_j is the *jth* principal factor and λ_{ij} corresponds to the factor loading of F_j on U_i. Only those factors which contribute significantly to the variance in the cluster are used to represent the cluster [77]. In-lines within the cluster are then identified with these principal factors. These in-lines, collectively known as *factor in-lines*, form the set of loosely correlated variables which are able to effectively represent the clusters to which they belong [78].

The data for problem decomposition is generated by running FABRICS repeatedly where each run is preceded by generating a realization of the disturbances. Each device simulation is followed by a circuit simulation using SPICE. The data is gathered in the form of vectors of process parameters, disturbances and their corresponding in-line, device and circuit performance parameter values. The clustering algorithm uses the n_u-dimensional vector of in-lines and creates a $n_u \times n_u$-dimensional correlation matrix. The matrix is symmetric and positive semi-definite. A search is then carried out over the upper triangular part of the matrix and an in-line is accepted into a cluster if its correlation coefficient with at least one of the in-lines already in the cluster is greater than ρ_T (which is determined based on the desired confidence level). If the in-line has a correlation coefficient less than ρ_T with all the in-lines in the existing clusters, it forms a new cluster. If, on the other hand, an in-line is a candidate for more than one cluster, then all such clusters are coalesced. Once the search is completed, the algorithm creates a correlation matrix for each of the clusters from the parent correlation matrix. These matrices then act as inputs to the factorization algorithm.

Using the constraint that factors of each cluster have to be chosen such that the first factor accounts for the maximum variance in the cluster, and all succeeding factors account for the maximum variance of the residual, the factor loadings can be represented in terms of the eigenvalues and eigenvectors of the correlation matrix $\mathbf{C}$ [42]. Specifically,

$$\lambda_{ji} = e_{ji} \frac{\sqrt{\Lambda_i}}{\sqrt{e_{1i}^2 + e_{2i}^2 + \cdots + e_{ni}^2}} \tag{6.3}$$

where

Λ_i

is the ith largest eigenvalue of **C** and

e_{ji}

is the jth element of the eigenvector corresponding to Λ_i.

The ratio of the ith eigenvalue to the trace of **C** represents the fraction of the variance accounted for by the ith principal component. Therefore all those eigenvalues which do not contribute significantly to the trace of the correlation matrix are deleted without loss of accuracy. Depending on the accuracy desired, the first k out of n eigenvalues are chosen. The eigensystem is determined by applying the method of *Upper Hessenberg Reduction* [74, 43] and **QR** decomposition to **C** repeatedly whereby the matrix is reduced to a diagonal form [37].

The algorithm then chooses the first k eigenvalues such that $\sum_{i=1}^{k} \Lambda_i / \sum_{i=1}^{n} \Lambda_i \geq \beta$ where β is the fraction of the variance specified by the user. It should be noted that the fractional error introduced by thresholding is $(1-\beta)$. Thus any value of ρ_T, used during clustering, which is less than $(1-\beta)$ will not cause any additional errors to be incurred due to misclassification. The corresponding eigenvectors are used to generate the loadings of the k principal factors using Eq. (6.3). The next step consists of identifying in-lines which are the closest approximations to the k chosen principal components. Since the cluster consists of inter-correlated in-lines, it is not possible to identify each of the factors with an unique in-line accurately. The criterion chosen for picking an in-line to represent a factor is that the factor must account for at least a certain minimum percentage of the in-line's variance (i.e. the factor will have to be strongly correlated to the in-line which will represent it). In case more than one in-line satisfies this condition, the one with the highest loading due to the factor is chosen. If no single in-line satisfies the correlation criterion, then a pair of loosely correlated in-lines which satisfy the criterion is chosen. Once the factor in-lines are derived all other in-lines in the cluster can be represented using them. However, since the non-factor in-lines are not necessarily observed in the same step as factor in-lines, they can be represented by the set of previous factor in-lines and the intervening process parameters and disturbances. The algorithm is biased towards choosing in-lines that are observed early in the process as factors. This is done by imposing a graded penalty for choosing in-lines observed late in the process. Biasing is favorable from the viewpoint of quality control since it is desirable to detect deviations in the process trajectory early in the process.

The results of the problem decomposition algorithm have been used to create models by the multi-layer regression software described in the next section. An example of the problem decomposition for a 4-device NMOS process is presented in conjunction with a device parameter model at the end of the next section.

6.4.2. Modeling for Process Control

The aim of the modeling subsystem is to create models of the fabrication process that are suitable for control, based on the data gathered from the fabrication line or simulators that emulate the fabrication process. The models relate the process parameters and disturbances to the in-lines, and the in-lines to the circuit performances. The total profit can be derived as a function of the *acceptability region* and the *jpdf* of the output performances. For effective control, it is desirable that the models be sufficiently accurate and efficient so that they can be repeatedly run for prediction as well as optimization.

The models are built using a multi-layer regression algorithm that determines the degree and coefficients of polynomials adaptively without paying the penalties of large CPU times and memory [45]. The implementation of this algorithm, MULREG, is described in [102]. Model building is performed in a sequence of layers. In each layer of multi-layer regression, a number of low-dimensional models are created using the sample data and intermediate variables are generated. Each intermediate variable is a polynomial of a subset of the total set of inputs to that layer and is the best possible representation of the output for that subset of inputs. Typically, such a subset will contain two inputs. The more significant of these intermediate variables, i.e. those which are better approximations to the output than the rest, serve as inputs to the next layer. Each layer is built by generating a number of low-dimensional models. Hence, unlike in the case of traditional regression methods, the additional memory required does not grow rapidly with an increase in the degree of the polynomial. The flowchart of MULREG can be found in Fig. 6-4.

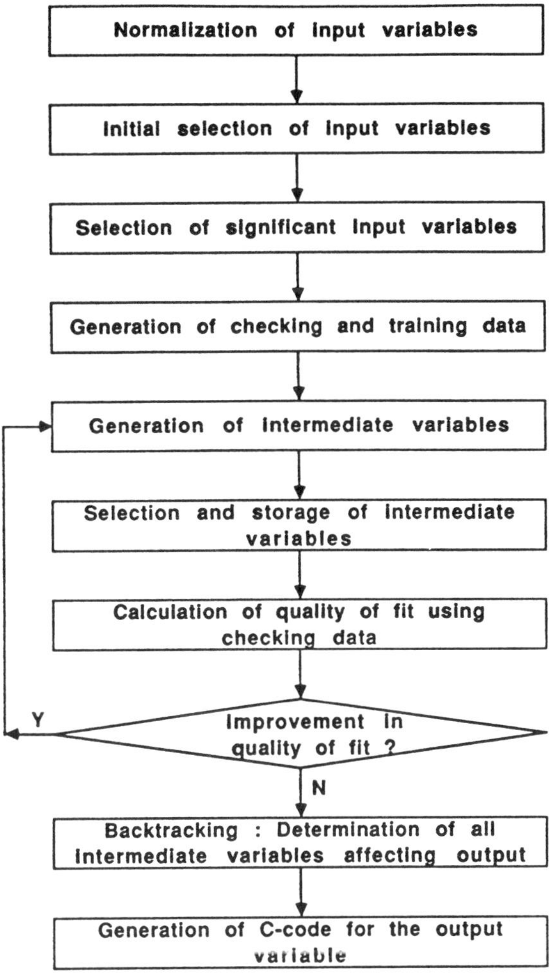

Figure 6-4: Flowchart of MULREG.

The factor in-lines in a cluster are represented by the process parameters and disturbances that govern the cluster while the non-factor in-lines and the device parameters/circuit performances are represented in terms of the factor in-lines as well as some of the process disturbances which affect the output performances but whose effects are not captured by the factor in-lines. Fig. 6-5 depicts an example of a two layer model of a device parameter created using MULREG. The device parameter (NE-CGBO) modeled is the gate to bulk capacitance of an enhancement transistor of a 4 device NMOS process, details of which are presented in Section 6. Data for the models was generated using FABRICS. The problem was decomposed into 10 subproblems and each of the device parameters generated by FABRICS was modeled. NE-CGBO was found to depend on two factor in-lines: thickness of gate oxide (*thsio2gate2*) and junction depth of the field implant (*xjf4*). The factor in-lines in turn were represented using 8 process parameters and disturbances, 3 for *thsio2gate2* and 5 for *xjf4*.

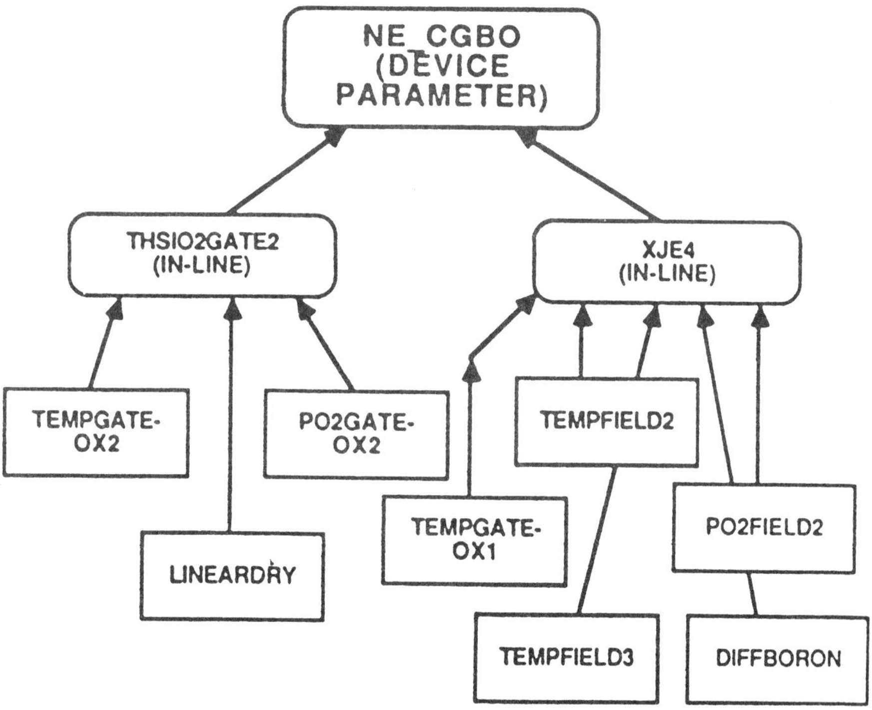

Figure 6-5: 2-layered model of a device parameter generated by the modeling software.

6.4.3. Statistical Quality Control

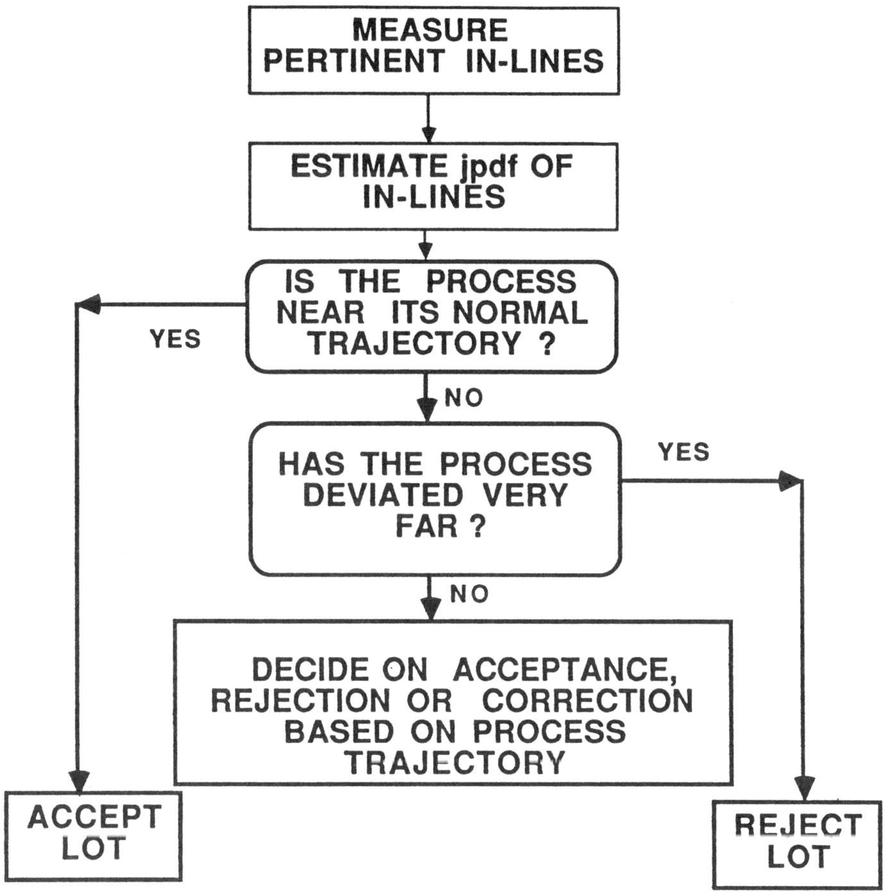

Figure 6-6: Quality control of an IC fabrication line.

The quality control subsystem (see Fig. 6-7) comprises several smaller software modules each of which indicates a decision made by the quality control subsystem. Together with these decision making modules are programs which indicate the activities that must precede the decision making process. These activities are determination of the process observables, estimation of the *joint probability density function* (*jpdf*) of the *observables* and determination of the *region of acceptability* based on

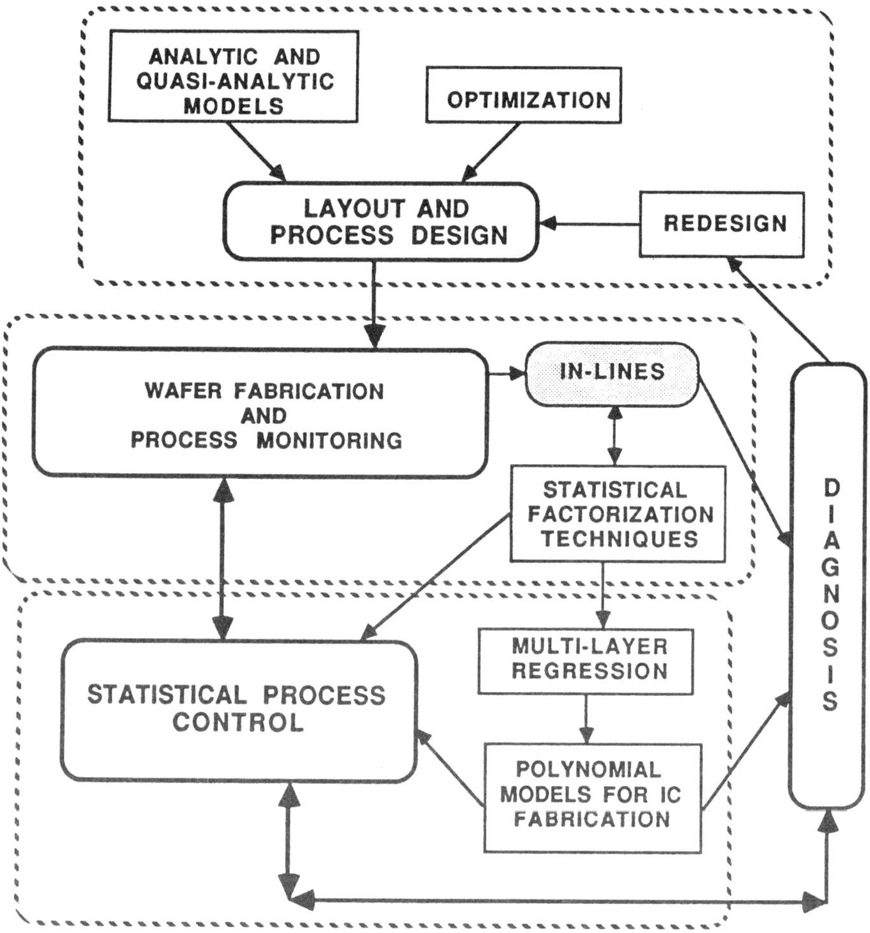

Figure 6-7: Overview of the control system.

the constraints specified by the customer. We first describe the algorithms that perform the above tasks and then the algorithms required to implement the three decision making systems.

The first step consists of determining the in-lines which will render the process observable, from the constraints and the objective. Process observability consists of identifying in-lines that affect the fabrication yield based on which the quality control decisions are made. The models created are appropriate for the identification. During problem decomposition quasi-independent factor in-lines are identified which act as inputs to the second layer of the model. Therefore the in-lines to be observed are a subset of the factor in-lines together with the process parameters and disturbances that form inputs to the second layer of the model. Since the models are polynomial in form, *dependency trees* which relate dependencies of the circuit performances on the factor in-lines, process parameters and disturbances can be identified by parsing the second layer. A dependency tree is a graph where the edges represent the model dependencies and the nodes represent the dependent parameters. Once the external constraints and the objective of interest are known, the system determines the circuit performances to which the objective is sensitive. Using the dependency trees the factor in-lines, process parameters and disturbances which are necessary to monitor these circuit performances are then determined. Throughout the remainder of this paper these parameters will be referred to as *observables* and will be represented by a vector of continuous random variables $\mathbf{W}$. The dependency trees can be extended similarly, from the viewpoint of feed forward control, to determine the process parameters which affect the output performances.

The next step consists of estimating the moments of the observables from the in-line measurements made during the process. The objective of interest for quality control, yield, is dependent on the *jpdf* of output performances, Ψ_Z. Since Ψ_Z in turn is dependent on the *jpdf* of the observables, Ψ_W, it is necessary to be able to statistically estimate Ψ_W. The estimation procedure comprises two parts, determining the type of the distribution, i.e. the moments which characterize the *jpdf*, and the sample size necessary to perform the estimation of the moments with the desired confidence. A simplification is made at this point by assuming that Ψ_W is multivariate normal. In our examples this assumption is not significantly erroneous. *Chi-square (goodness of fit) tests* show that most of the in-lines display a trend towards normality while most of the remaining can be accounted for as log-normal (e.g. source/drain sheet resistance in an NMOS process). All the disturbances are assumed to be normally distributed and independent. The resultant *jpdf* Ψ_W in the space of observables, which constitute a n_w-dimensional space, can be represented in terms of a mean vector $E[\mathbf{W}] = (E[W_1], \cdots ,E[W_{n_w}])$ and a covariance matrix Σ_W, and expressed as

$$\Psi_W = \frac{1}{[(2\Pi)^{n_w} \det \Sigma_W]^{1/2}} exp(-\frac{1}{2}[\mathbf{W}-E[\mathbf{W}]]^T \Sigma_W^{-1} [\mathbf{W}-E[\mathbf{W}]]). \qquad (6.4)$$

The problem of estimating the Ψ_W then reduces to the problem of estimating $E[\mathbf{W}]$ and Σ_W. The sample size required for the estimation, which in turn determines the number of in-line measurements, is dependent on the number of moments that need to be estimated and the confidence desired. Expressions relating the confidence interval in terms of the sample size and the order of the moments exist in literature [46]. Once the confidence interval is decided upon, the sample size can be calculated. In general, the increase in the sample size is superlinear for a linear decrease in the confidence interval. A minimum sample size for each of the parameters, mean, variance and correlation coefficient, is determined from the corresponding confidence intervals. The maximum of the three is then chosen as the required sample size. In practice the sample size is also restricted by the practical cost limitations on the number of test structures that can be used and the measurements that can be made in a fabrication process. A tradeoff between the cost and the accuracy desired (reflected by the confidence interval) has to be used to derive the sample size.

The third part of the quality control system deals with the determination of the acceptability region in the space of observables. The need for determination of the acceptability region stems from the fact that the acceptability region specified by the customer is in the space of circuit performances while the in-line measurements are performed to estimate the *jpdf* in the space of observables. The deviation of the process from its normal trajectory is determined by comparing the *jpdf*s of the normal and the observed processes. The constraints in conjunction with the process deviation enable the system to make one of the four quality control decisions. Two distinct possibilities exist

1. To evaluate Ψ_Z from the estimated Ψ_W using the models.

2. To derive the acceptability region in the in-line space, $\mathbf{R_W}$, from the acceptability region in the circuit performance space, $\mathbf{R_Z}$. This can be achieved by the method of *Simplicial Approximation* [23].

If Ψ_Z is to be estimated from Ψ_W then either a mapping of the moments of **Z** in terms of the moments of **W** must be derived from the existing models, or a point by point mapping of the samples in the **W** space to samples in the **Z** space must be done. The disadvantage of this method is that it requires a knowledge of the type of distribution of Ψ_Z. In addition, mapping the samples from one space to the other would be extremely time consuming. From the viewpoint of static quality control the inverse mapping of the acceptability region has to be done only once at the start of the process. It needs to be changed only if the process model is changed. Such a change may be required when a process parameter is significantly modified as a result of either feed forward control or process diagnosis. The mapping of the acceptability region in the space of performance parameters, $\mathbf{R}_Z$, to the acceptability region in the space of observables, $\mathbf{R}_W$, cannot be performed using a point to point mapping scheme. This is because the complexity and non-linearity of the models make the derivation of an inverse mapping infeasible. Therefore the system uses simplicial approximation as the method of deriving the acceptability region in the space of observables. Simplicial approximation uses an exact acceptability region in the output space and the forward transformation to generate an interior approximation of the acceptability region in the input space.

The method of simplicial approximation locates and approximates the boundary of an acceptability region in an n-dimensional input space by a polyhedron of bounding simplices given the corresponding acceptability region in the output space and a mapping from the input to the output space [23]. The algorithm for simplicial approximation can be found in [23, 77]. Once the simplicial approximation to the acceptability region in the **W** space is determined, the quality control decision criteria can be derived based upon the estimated Ψ_W.

6.4.4. Acceptance and Rejection Criteria

The output from the systems described in the previous section is used by the decision-making part of the quality control system. Decision making consists of a sequence of evaluation steps that are based on the deviation of the observed *jpdf* from that of the nominal process. The procedure for rejection is based on identifying cases where the process deviation is large enough to be detected via a single in-line or a linear combination of them. Acceptance on the other hand is based on cases where the *jpdf*s are not significantly *different*. In this section we will describe the algorithms for rejection and acceptance that are inexpensive compared to the direct approach - estimating the fabrication yield and then deciding on acceptance, rejection or correction.

The difficulty associated with the estimation of yield by integration is due to the complexity of the indicator function, which is reflected in the complexity and number of constraint equations that define the acceptability region in the space of observables. The indicator function in the space of observables (Φ_W) is expressed as

$$\Phi(W) = \begin{cases} 1, & W \subset R_W \\ 0, & \text{otherwise} \end{cases} .$$

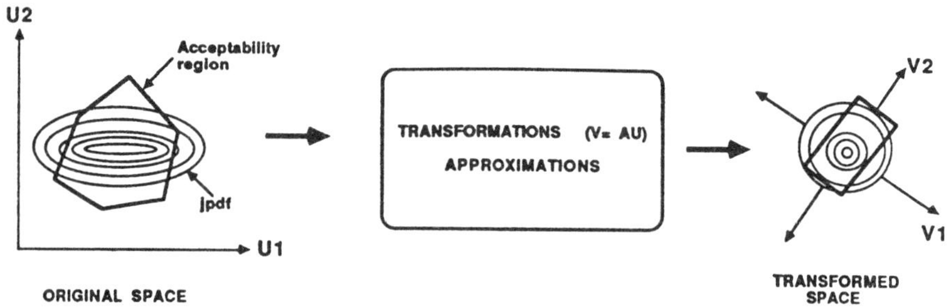

Figure 6-8: Yield estimation.

The problem of a complex indicator function is overcome in the quality control system by approximating the acceptability region by a hyperbox. The integrals can then be decoupled and the final yield can be approximated as the product of several partial yields (see Fig. 6-8). Using the approximation that the *jpdf* in the in-line space is multivariate normal and that the various factor in-lines are independent Σ_W and Σ_W^{-1} in Eq. (6.4) are diagonal. Therefore, if the acceptability region is a hyperbox the yield can be expressed as

$$\int_{-\infty}^{\infty} \cdots \int_{-\infty}^{\infty} \Psi_W(\mathbf{W})\Phi_W(\mathbf{W})dw_1 \cdots dw_{nw}$$

$$= \int_{L_1}^{H_1} p(w_1)dw_1 \cdots \int_{L_{nw}}^{H_{nw}} p(w_{nw})dw_{nw} \tag{6.5}$$

where

L_i

denotes the lower bound of the hyperbox in the ith dimension,

H_i

denotes the upper bound of the hyperbox in the ith dimension,

$p(w_i)$

denotes the marginal distribution of W_i.

The orientation of the hyperbox is derived using a heuristic. The *largest face* of the acceptability region, which is determined by the simplicial approximation algorithm, is aligned with one of the faces of the hyperbox. Of the other (2n-1) faces, (2n-2) are orthogonal and one is parallel to the first face. These faces pass through the extreme points of the polytope in the direction of outward normal to the planes. Determination of the first plane also defines exactly the rotation necessary for the coordinate axes to align the box to the new coordinate system. Let this rotation operation be denoted by a matrix **A** representing a linear transformation of the original coordinate axes to a new set of axes. Given that the transformation of the axes is expressed as

$$\mathbf{V} = \mathbf{AW} \tag{6.6}$$

the *jpdf* in the **V** space, Ψ_V, can be expressed as a multivariate normal distribution similar to Eq. (6.4), where $\Sigma_V = \mathbf{A}\Sigma_W\mathbf{A}^T$ and $E[\mathbf{V}] = \mathbf{A}E[\mathbf{W}]$.

Since the transformation corresponds to pure rotation, **A** is unitary. Matrix **A** can be derived using the idempotency and conjugacy properties of solutions to orthonormal transformations. However, even when **A** is unitary Σ_V is not diagonal in general, which means that decoupling cannot be achieved. The situation is remedied in the system by scaling the original space so that the covariance matrix is an identity. If the scaling matrix is denoted by **S**, where **S** is diagonal with $S_{ii} = 1/\sigma_{ii}$, then the resultant transformation can be expressed as

$$\mathbf{V} = \mathbf{Q}\mathbf{W} \tag{6.7}$$

where $\mathbf{Q} = \mathbf{AS}$. The resulting Ψ_V has a diagonal covariance matrix $\Sigma_V = \mathbf{Q}\Sigma_W\mathbf{Q}^T = \mathbf{I}$, and the integral can be expressed as a product of n_v integrals

$$Y_V = \prod_{1}^{n_v} y_i \tag{6.8}$$

$$\text{where} \quad y_i = \int_{L_i}^{H_i} p(v_i)d(v_i).$$

The limits of integration, L_i and H_i, correspond to the equations of the hyperbox in the transformed space. The criterion of rejection in terms of the ith partial yield can be written as

$$y_i \leq \frac{Y_{rej}}{\prod_{j=1, j \neq i}^{n_v} y_j} \tag{6.9}$$

where Y_{rej} represents the rejection threshold.

To keep the decision making process tractable, it is assumed that the process trajectory variation leaves the type of distribution unchanged, i.e., the new distribution can be approximated by a multivariate normal. The system at the start of the process assumes that the *jpdf* of the in-lines to be observed in the present process is the same as that of the normal process. As the process proceeds sequentially, in-lines are measured and the marginal distributions are updated. The marginal distributions are then translated to attain a zero mean. Since this is only a translation, the matrix **A** remains unchanged, while matrix **S** is changed according to the estimated σ_i. The new **V** space would have the same *jpdf* as before with a zero mean vector, while the hyperbox would have shifted in the direction of W_i thereby generating a new set of H_i and L_i in Eq. (6.8).

The system is also geared to decide on rejection based on the change in moments of a single in-line. For this purpose the system analyzes effect of mean and variance changes separately. Using the assumption of normality, the partial yield y_i can be expressed as a difference of two *error functions*

$$y_i = erf(H_i) - erf(L_i)\,. \tag{6.10}$$

Using the polynomial expansion of the error function, the partial yield y_i can be reduced to a power series in H_i and L_i. Depending on the accuracy desired the series can be truncated to m terms. Therefore the effect of a mean shift on the yield, without changing the variance, can be written as

$$y_i = \sum_{j=1}^{m} \frac{(-1)^j (H_i - \Delta E[V_i])^{2j+1}}{j!(2j+1)} - \sum_{j=1}^{m} \frac{(-1)^j (L_i - \Delta E[V_i])^{2j+1}}{j!(2j+1)} \tag{6.11}$$

where $\Delta E[V_i]$ is the shift in mean in the V_i dimension.
The bounds on $\Delta E[V_i]$ are determined by solving for the roots of the polynomial where y_i satisfies the equality condition in Eq. (6.9). Similar equations are solved for bounds in all the n_v dimensions. Whenever a marginal distribution is estimated, the mean shift $\Delta E[\mathbf{V}]$ is derived in terms of $\Delta E[\mathbf{W}]$ using the transformation given by Eq. (6.7). If the observed shift in mean of the marginal distribution of W_i is $\Delta W\mu_i$, the corresponding shift in the $\mathbf{V}$ space is

$$\Delta E[\mathbf{V}] = \mathbf{Q}_i \Delta E[W_i]. \tag{6.12}$$

If the detected mean shift in any dimension of $\mathbf{V}$ exceeds that of its corresponding bound the system decides to reject the lot from further processing.

The effect of a variance increase in the $\mathbf{W}$ space is observed as a scaling down of the hyperbox in the $\mathbf{V}$ space. Therefore the corresponding equation for the bound on the variance is expressed as

$$y_i = \sum_{j=1}^{m} \frac{(-1)^j H_i^{2j+1}}{j!(2j+1)S_i} - \sum_{j=1}^{m} \frac{(-1)^j L_i^{2j+1}}{j!(2j+1)S_i} \tag{6.13}$$

where S_i is termed *critical scale factor* and is the ratio of the estimated variance to the variance observed during the normal process of the marginal distribution of V_i. During the process, if the system detects that the bounds on one or more dimension of $\mathbf{V}$ are scaled by factors greater than their corresponding critical scale factors, then the lot is rejected as in the case of large mean shifts.

The system uses as its criterion for acceptance a measure of the difference between the observed and expected *jpdfs* and its relation to the final objective, yield. Using a discriminant function which converts the multivariate populations into univariate populations, the means of the populations can be separated as far as possible with respect to the population covariance. The function used by the system is the *Fisher's linear discriminant function* [46] , and the resulting distance is the *Mahalanobis distance* [95] expressed as

$$M = \Delta\mu^T \Sigma^{-1} \Delta\mu \tag{6.14}$$

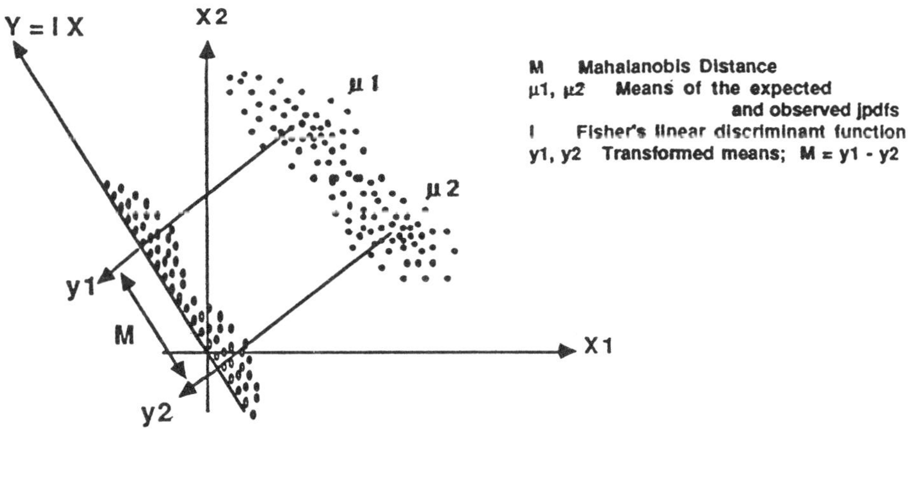

Figure 6-9: Graphic representation of the Mahalanobis distance.

where $\Delta\mu$ is the difference between the mean vectors (see Fig. 6-9). The distance is calculated under the assumption that the covariance matrix Σ is the same for both populations. Since there are slight deviations, a pooled estimate of the covariance matrix is used in practice. The change in the yield is then related to the distance by weighting it with the yield sensitivity in the direction of the shift. This is equivalent to weighting the shift in mean along each dimension of **W** by the sensitivity of the yield to the corresponding in-line. The *weighted Mahalanobis distance* is expressed as

$$M_S = \Delta\mu^T \mathbf{Y}_s{}^T \Sigma^{-1} \mathbf{Y}_s \Delta\mu \tag{6.15}$$

where $\mathbf{Y}_s$ is a diagonal matrix with $y_{sii} = \partial Y / \partial W_i$. The weighted distance is used by the system as a measure of yield fluctuations due to small changes in the mean. The method can be applied for any objective function given the sensitivities of the W_i's to the objective. At present we are using the yield as the objective and the methods of determining the yield gradients can be found in [117, 102]. If the yield change is smaller than the difference between the expected yield and the *threshold of acceptance* the system decides on acceptance without any further steps.

When none of these conclusions can be drawn with sufficient confidence, the system predicts the yield using the acceptability region derived by simplicial approximation in the space of observables. The decisions to accept or reject are then made based on this estimated yield. Under certain conditions, especially when the predicted yield is not far below the threshold of acceptance corrective measures are taken by the system by manipulating the subsequent process parameters for the lot under observation. The following section describes this feed forward control system.

6.4.5. Feed Forward Control

The aim of feed forward control is to improve lot yield by modifying the process conditions for the subsequent stages. Before this step can be carried out, the current state of the fabrication process has to be identified. This is the first stage of feed forward control which consists of process identification, i.e., determination of the distributions of the process disturbances of all the previous stages in the fabrication process. This stage is followed by iterative yield improvement which requires the calculation of the yield derivatives with respect to the future process control parameters. In order to reduce the dimensionality of the feed forward control problem, the circuit performances are partitioned into separate groups which are very loosely correlated. A method similar to the one used to derive in-line clusters is used to group the circuit performances. Each group of circuit performances is affected by process control parameters, layout parameters and disturbances which do not affect any other group. The feed forward control procedure can be individually applied to each of these groups.

The models relating the in-line measurements to the input parameters are used for process identification. The process disturbances are all independent random variables. In this paper, we have assumed normal distributions for all the disturbances. The problem of identification is now reduced to one of determining the means and standard deviations of all the disturbances in the previous stages.

The disturbances are of two kinds - some affect the process control or layout parameters directly and others affect them indirectly. The process control parameter affects the mean of the first kind of disturbance. Let us consider the process control parameter, temperature of oxidation. We may nominally want it to be 950 °C. However, due to thermocouple calibration error (a disturbance), the mean of the distribution of the temperature could be 955 °C. Therefore, the mean of the process disturbance would now be 5 °C. The standard deviation would be non-zero since all the wafers in a lot would not experience the same temperature in lieu of their relative positions with respect to the heat source. The mean of this kind of disturbance thus has two components, μ_d (value of the process control parameter) and μ_r (random) and its standard deviation would have a single component, σ_r (random). Thus, the probability density function, $f(s)$, of such a disturbance can be written as,

$$f(s) = \frac{1}{\sqrt{2\pi}\sigma_r} exp(- \frac{(s - \mu_d - \mu_r)^2}{2\sigma_r^2}). \tag{6.16}$$

The second kind of process disturbance has a mean, μ_r and a standard deviation, σ_r. It does not have a process control parameter contributing to its mean. Thus, the probability density function, $f(s)$ can be written as,

$$f(s) = \frac{1}{\sqrt{2\pi}\sigma_r} exp(-\frac{(s-\mu_r)^2}{2\sigma_r^2}). \tag{6.17}$$

Eq. (6.17) can be considered to be a special case of Eq. (6.16) with $\mu_d = 0$. Typically, for a fabrication line, the nominal distributions of the process disturbances would be known. However, due to the fluctuations mentioned earlier, the means and standard deviations of their distributions could have been modified.

The procedure for determining the means and standard deviations is described in [105]. First, from the in-line measurements, the means of all the in-lines are determined. Then, the *Han-Powell* [92] variable metric nonlinear constrained minimization algorithm with the *watchdog technique* [17] is used to solve this minimization problem.

While attempting to determine the subsequent process control parameters, we will assume nominal distributions for all the future disturbances. Let there be N_i process disturbances in the *ith* group. Let the deterministic component of the means for the disturbances (i.e. process control parameters[2]) be $\mu_{d1}, \mu_{d2}, \ldots \mu_{dN_i}$, the random component of their means be $\mu_{r1}, \mu_{r2}, \ldots \mu_{rN_i}$ and their standard deviations be $\sigma_1, \sigma_2, \ldots \sigma_{N_i}$. Let s_i refer to any point in the *ith* subspace of input parameters and let s_{ij} be its *jth* coordinate. Since, the process disturbances are statistically independent, the jpdf will be,

$$\Psi_i(s_i) = \prod_{j=1}^{N_i} \frac{1}{\sqrt{2\pi}\sigma_j} exp(-\frac{(s_{ij}-\mu_{dj}-\mu_{rj})^2}{2\sigma_j^2}). \tag{6.18}$$

The *ith* partial yield Y_i is given by,

$$Y_i = \int_{s_i} \Psi_i(s_i)\phi_i(s_i)ds_i \tag{6.19}$$

[2]For the second kind of process disturbance this will be zero.

where

$$\phi_i(s_i) = \begin{cases} 1 & \text{if } s_i \text{ is within the acceptability region} \\ 0 & \text{otherwise} \end{cases} \tag{6.20}$$

The parameters that can be varied are the process control parameters which contribute to the means of the process disturbances. The derivatives of the yield with respect to each of these have to be determined. Since, only the *ith* partial yield depends on μ_{dj}, the product of the remaining partial yields can be treated as a constant, Q_i, where $Y_i Q_i = Y$. The yield derivative with respect to μ_{dj} is given by,

$$\Delta_j = Q_i \frac{\delta}{\delta \mu_{dj}} \int_{s_i} \Psi_i(s_i)\phi_i(s_i)ds_i. \tag{6.21}$$

Since the integration is with respect to s_i and the differentiation is with respect to μ_{dj}, the order in which they are performed can be changed. Further, since only $\Psi_i(s_i)$ depends on μ_{dj},

$$\Delta_j = Q_i \int_{s_i} \frac{\delta \Psi_i(s_i)}{\delta \mu_{dj}} \phi(s_i)ds_i. \tag{6.22}$$

Therefore,

$$\Delta_j = \frac{Q_i}{\sigma_j^2} \int_{s_i} \Psi_i(s_i)\phi_i(s_i)(s_{ij} - \mu_{dj} - \mu_{rj})ds_i. \tag{6.23}$$

In [117], it has been suggested that yield gradient methods can be used for yield optimization at the design stage. We have extended the idea to include yield optimization at the fabrication stage also. Yield prediction during the fabrication process is discussed in [70]. The algorithm for determining the new set of values for the future process control parameters such that the estimated final yield is above an acceptable value is presented in the form of a flowchart in Fig. 6-10.

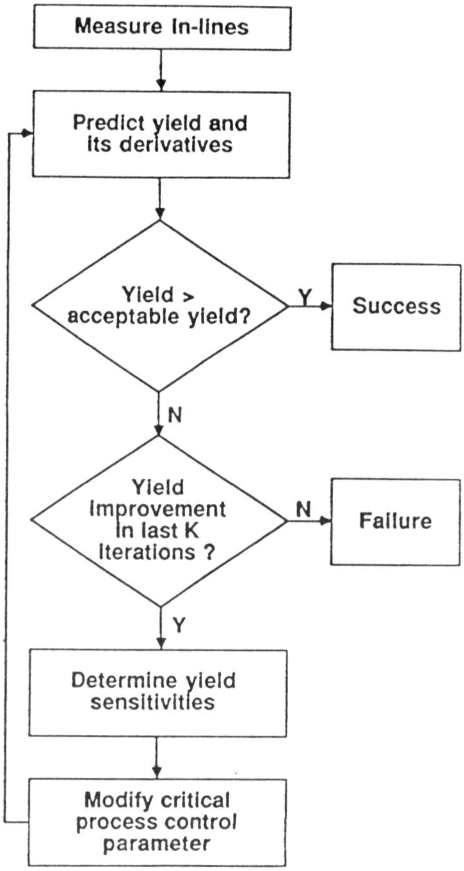

Figure 6-10: Flowchart of feed forward control algorithm.

6.5. Computational Examples

In this section we present computational examples to illustrate the software implementation of the various algorithms that have been described in the previous sections. We divide our description of the software system into three parts: modeling, quality control and feed forward control. The data flow in the software is graphically depicted in Fig. 6-11. All the programs are written in C and run under the UNIX operating system on a VAX-11/785. FABRICS was tuned to a commercial 4-device **NMOS** process, the devices being strong depletion, strong enhancement, weak depletion and weak enhancement. A large sample size was chosen and FABRICS and SPICE run to generate the data in form of vectors that could be used both by the problem decomposition as well as the regression routines.

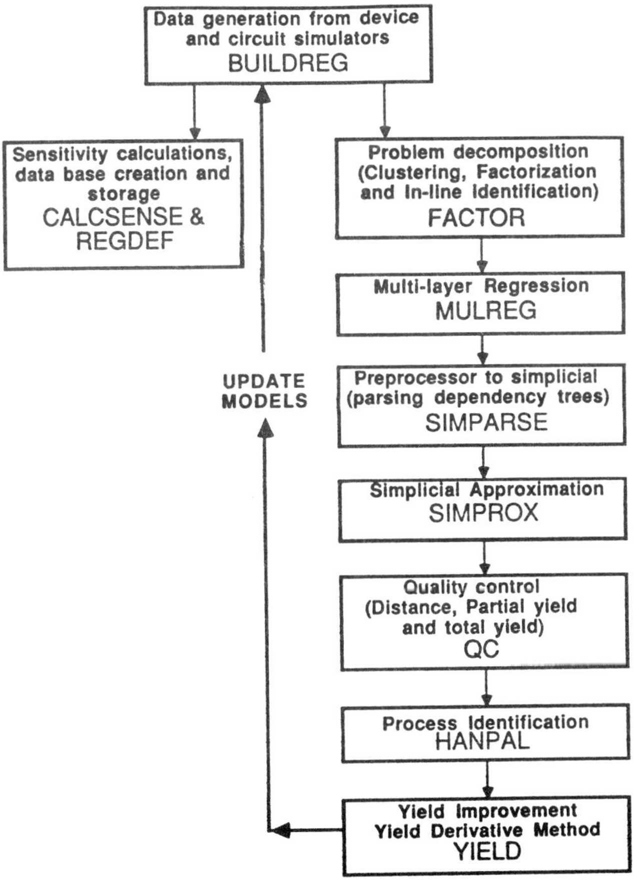

Figure 6-11: Overview of CMU-CAM system software.

The flowchart of the modeling software is depicted in Fig. 6-12 and the

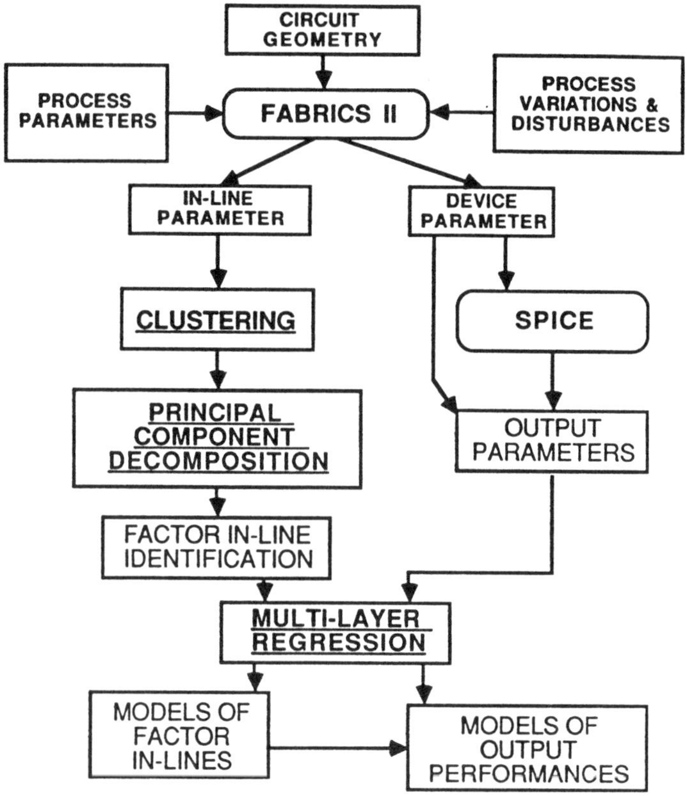

Figure 6-12: Overview of modeling software.

flowchart of the entire software system is given in Fig. 6-11. The sensitivities of the output parameters to the disturbances are calculated by a program called CALCSENSE. The data grouping and vectorization is done by a program called BUILDREG. BUILDREG accepts the user specified sample size as input and produces the necessary vectors in ascii or binary format. Problem decomposition is achieved by the program FACTOR. This routine needs the intra-cluster correlation value and the set of in-line vectors created by BUILDREG. FACTOR parses the input files, derives the correlation matrix, does the clustering and computes the correlation matrix for each of them.

Using $\rho_T = 0.1$, 10 clusters were formed. Experimentation showed that the clusters remained invariant when ρ_T varied between 0.08 and 0.17. For the principal component decomposition and identification part, a variance cutoff value (β) of 0.975 and a penalty value of 0.98 were used. The penalty of 0.98 indicates that if the jth in-line in a cluster is chosen instead of the ith in-line to represent a principal component, where $j \geq i$, then the loading on j, due to the principal component, must be greater by a factor of $(0.98)^{i-j}$. Most of the principal components could be represented effectively by a single in-line and the remaining few by a pair.

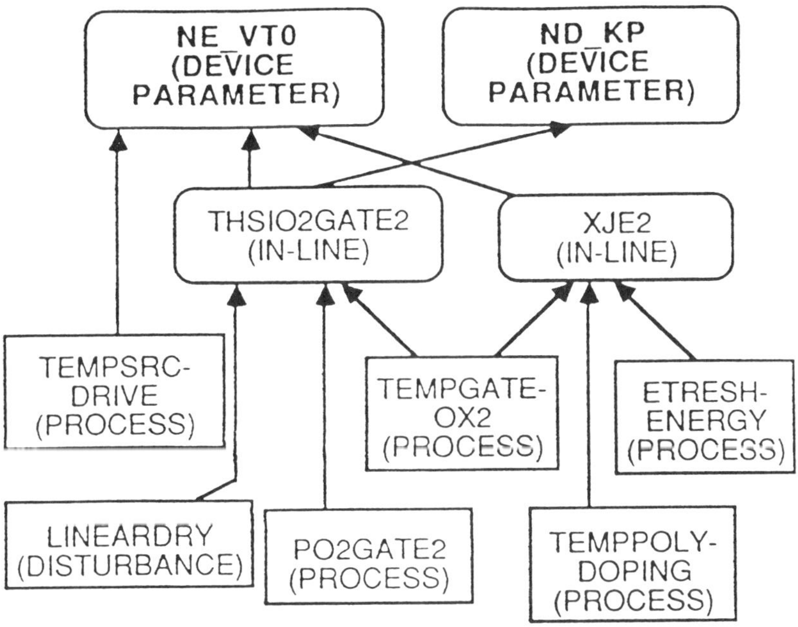

Figure 6-13: Example of a 2-layered model.

An example of the process model can be found in Fig. 6-13. The device parameters modeled here are the zero bias threshold voltage of the strong enhancement device, NE-VT0, and the transconductance of the strong depletion device, ND-KP. It is observed that the device parameters are dependent on two factor in-lines and a process parameter. The two factor in-lines are *thsio2gate2* (gate oxide thickness after the second gate oxidation step) and *xje2* (depth of the enhancement threshold implant after poly doping), and the process parameter is *tempsourcedrive* (temperature of source drive in). *Tempsourcedrive* changes the source and drain profiles thereby changing the gate controlled charge and hence the threshold voltage. Both the factor in-lines are observed before the source-drain implantation is performed and hence it was necessary to include the process parameter *tempsourcedrive*. The factor in-line *thsio2gate2* depends on two process parameters *tempgateox2* (temperature of second gate oxidation) and *po2gateox2* (partial pressure of oxygen in second gate oxidation), and one disturbance, *lineardry* (linear coefficient of dry oxidation), while *xje2* depends on three process parameters namely *Ethreshenergy* (the energy of the strong threshold implant), *temppolydoping* (temperature of the poly doping step) and *tempgateox2* (temperature of the second gate oxidation step). The model is pseudo-two layer due to the inclusion of a process parameter in the second layer. One of the reasons for this can be the failure to observe in-lines which accommodate the effect of variations in *tempsourcedrive*. However, it is also possible that the variation in in-lines due to *tempsourcedrive* are not major, in which case they do not contribute largely to the variance of the cluster, to which they belong, and are hence eliminated during problem decomposition.

The programs SIMPARSE, SIMPROX and MARGIN form part of the quality control system. SIMPARSE parses the models formed by MULREG and creates the dependency trees necessary for SIMPROX to determine the acceptability region in the space of observables using simplicial approximation. SIMPROX uses the dependency trees and the constraints on the circuit performances to determine the subset of observables that form the effective input space in terms of which the acceptability region is to be determined. The acceptability region is specified in terms of equations of hyperplanes. The final acceptability region is specified by generating a C procedure which consists of the equations of the hyperplanes in the form of inequalities. MARGIN uses the sample points generated by BUILDREG and determines the *Mahalanobis distance* as well as the partial yields. In case the user wants to determine the accurate yield or a decision cannot be made on the basis of acceptance or rejection criteria, MARGIN uses the C procedure generated by SIMPROX and computes the fabrication yield.

We used FABRICS to simulate a process with a small change in the moments of the process disturbances. A tolerance of 20% was allowed on

NE-VT0 and 10% on ND-KP. The yield for the normal process was approximately 85%. When the mean values of the process parameters were slightly perturbed the results obtained from MARGIN were:

Partial yield due to thsio2gate2 = 0.793
Partial yield due to xje2 = 0.983
Partial yield due tempsourcedrive = 0.928
Mahalanobis distance = 0.0267
The estimated fabrication yield = 71.0%

The high value of the Mahalanobis distance indicates a need for feed forward control. This is confirmed by the low value of the estimated parametric yield.

6.5.1. Yield Enhancement

In this section we will describe how the program YIELD was used after the gate oxidation stage of the fabrication process (simulated using FABRICS) to modify the process control parameters of the succeeding stages when the disturbances of the previous stages were shifted from their normal values. In the example described above, the estimated fabrication yield of 71% is below the threshold of acceptance. The feed forward control procedure can be invoked to increase the yield to a value above 80% (acceptable value - computed using the profit function). Given below is the list of process control parameters and disturbances that affect the device parameters of interest (see Fig. 6-13) together with their nominal values.

Name	Deterministic part of mean	Non-deterministic part of mean	Variance
tempgateox2 [$^\circ$C]	1000.0	1.0	2.0
po2gateox2 [-]	0.970	0.002	0.002
Ethreshenergy [eV]	110.0	2.0	1.0
temppolydoping [$^\circ$C]	950.0	-1.0	2.0
tempsourcedrive [$^\circ$C]	1000.0	-3.0	2.0
lineardry [μm/sec]	0.0	1840.0	20.0

Using the program, YIELD, the yield for this nominal set of values for the process control parameters and disturbances was found to be around 85%. This took 45 minutes of CPU time on the microvax (VAXstation II/GPX).

The process control parameters and disturbances affecting the gate oxide were found to be shifted from their nominal values. The new values for these are given below.

Name	Deterministic part of mean	Non-deterministic part of mean	Variance
tempgateox2 [°C]	1000.0	-4.0	2.0
po2gateox2 [-]	0.970	0.005	0.002
lineardry [μm/sec]	0.0	1875.0	25.0

The program, YIELD, was now used to modify the future process control parameters so that the expected final yield was greater than the acceptable value. From an initial value of 71%, the program was able to increase the expected value of the final yield above 80% by modifying just one process control parameter. This took 2 hours of CPU time on the microvax. The final distributions for the future process control parameters are given below.

Name	Deterministic part of mean	Non-deterministic part of mean	Variance
Ethreshenergy [eV]	110.0	2.0	1.0
temppolydoping [°C]	950.0	-1.0	2.0
tempsourcedrive [°C]	997.0	-3.0	2.0

Below, we briefly present two more examples. In the first, the output parameter is NE-CGBO (gate-bulk capacitance). The observables are thsio2gate2 (thickness of gate oxide) and xjf4 (field junction depth).

Acceptable yield = 85%
Mahalanobis distance between the two distributions = 6.3005e-03
Partial yield due to xjf4 = 99%
Partial yield due to thsio2gate2 = 93.8%
Estimated yield is = 93.0%

The small Mahalanobis distance indicates that this is a clear case of acceptance. None of the partial yields is low enough to indicate that the lot has to be rejected. This is verified by estimating the final yield (93%).

In the next example, the output parameter is NE-CGSO (gate-source overlap capacitance). The only observable is thsio2gate2. Hence, the partial yield due to thsio2gate2 is also the estimated final yield.

Acceptable yield = 85%
Partial yield due to thsio2gate2 = 77%
Mahalanobis distance between the two distributions = 9.7272e-02

Since the partial yield is lower than the acceptable value, feed forward control is required. The Mahalanobis is also high, indicating that the process has strayed significantly from its normal trajectory.

Nominal values for the input parameters:

Name	Deterministic part of mean	Non-deterministic part of mean	Variance
tempgateox2 [°C]	1000.0	1.0	2.0
po2gateox2 [-]	0.970	0.002	0.002
lineardry [μm/sec]	0.0	1840.0	25.0
tempsourcedrive [°C]	995.0	2.0	2.0

Values of inputs after gate oxidation:

Name	Deterministic part of mean	Non-deterministic part of mean	Variance
tempgateox2 [°C]	1000.0	-5.0	2.0
po2gateox2 [-]	0.970	0.005	0.002
lineardry [μm/sec]	0.0	1868.0	25.0

By using the program YIELD, we determined that the estimated yield can be increased from 77% to 85% by decreasing tempsourcedrive from 995 °C to 990 °C.

6.6. Conclusions

A portable software system which performs process modeling, quality control and feed forward control has been described in this paper. Data needed for this system is readily available in any presently installed CAM system (such as COMETS or PROMIS). The performance of our system, viewed in terms of the total CPU time as a fraction of the total processing time, is quite satisfactory for the typical process flow in IC manufacturing.

References

[1] Y. Akasaka and K. Horie.
Lateral Spread of Boron Ions Implanted in Silicon.
Appl. Phys. Lett. 21(4), 1982.

[2] D.A. Antoniadis, S.E. Hansen, and R.W. Dutton.
SUPREM II - A Program for IC Process Modeling and Simulation.
Technical Report 5019-2, Integrated Circuit Laboratory, Stanford University, June, 1978.

[3] K.J. Antreich and R.K. Koblitz.
Design Centering by Yield Prediction.
IEEE Trans. on Circuits and Systems CAS-29(2):88-95, February, 1982.

[4] Y.Aoki, T.Toyabe,S.Asai and T.Hagiwara.
CASTAM: A Process Variation Analysis Simulator for MOS LSI's.
IEEE Trans. on Electron Devices ED-31:1462-1467, 1984.

[5] P.Balaban and J.J.Golembeski.
Statistical Analysis for Practical Circuit Design.
IEEE Trans. on Circuit and Systems CAS-22:100-108, Feb., 1975.

[6] J.W. Bandler, P.C. Liu, and J.W. Chen.
Worst Case Network Tolerance Optimization.
IEEE Transactions on Microwave Theory and Techniques MTT-23(8):630-640, August, 1975.

[7] J.W. Bandler and H.L. Abdel-Malek.
Optimal Centering, Tolerancing, and Yield Determination via Updated Approximations and Cuts.
IEEE Trans. on Circuits and Systems CAS-25(10):853-871, October, 1978.

[8] J.L.Bentley, D.Haken and R.W.Hon.
Statistics on VLSI Designs.
Technical Report, Carnegie-Mellon University, Department of Computer Science, April, 1982.

[9] J. Bernard.
The IC Yield Problem: A Tentative Analysis for MOS/SOS Circuits.
IEEE Trans. on Electron Devices ED-25:939-944, August, 1978.

[10] G. E. Box, W. G. Hunter, J. S. Hunter.
Statistics for Experimenters.
J. Wiley - Interscience, 1978.

[11] G. E. P. Box.
Studies in Quality Improvement: Signal to Noise Ratios, Performance Criteria and Statistical Analysis.
Technical Report 11, center for Quality and Productivity Improvement, University of Wisconsin-Madison, March, 1986.

[12] R. K. Brayton, C. L. Chen, R. J. H. M. Otten and J. Y. Yamour.
A Silicon Compiler for Automated Custom Macro Design.
In *Proc. of IEEE Custom IC Conference.* Rochester, May, 1984.

[13] R.K. Brayton, S.W. Director and G.D. Hachtel.
Yield Maximization and Worst-Case Design with Arbitrary Statistical Distributions.
IEEE Trans. on Circuits and Systems CAS-27(9):756-764, September, 1980.

[14] D. K. Brice.
Ion Implantation Range and Energy Deposition Distributions.
IFI/PLENUM, 1975.

[15] R.L. Burden and J.D. Faires.
: *Numerical Analysis (3rd ed.).*
Prindle, Weber & Schmidt, Boston, 1985.

[16] E.M. Butler.
Realistic Design Using Large-Change Sensitivities and Performance Contours.
IEEE Tansactions on Circuit Theory CT-18(1):58-65, January, 1971.

[17] R. M. Chamberlain, M. J. D. Powell, C. Lemarechal and H. C. Pedersen.
The Watchdog Technique for Forcing Convergence in Algorithms for Constrained Optimization.
Proceedings of the 10th International Symposium on Mathematical Programming, Montreal , 1979.

[18] I. Chen and A.J. Strojwas.
Realistic Yield Simulation for VLSIC Structural Failures.
IEEE Transactions on CAD of ICAS CAD-6(6):965-980, 1987.

[19] M. Chew and A. J. Strojwas.
Efficient Circuit Re-extraction for Yield Simulation Applications.
In *IEEE International Conference on Computer-Aided Design (ICCAD) Digest of Technical Papers*. IEEE, November, 1987.

[20] P. Cox, P. Yang and P. Chatterjee.
Statistical Modeling for Efficient Parametric Yield Estimation.
In *Proc. of International Electron Device Meeting*, pages 242-245. 1984.

[21] D.C. D'Avanzo M. Vanzi, and R.W. Dutton.
One-Dimensional Semiconductor Device Analysis (SEDAN).
Tech. Rep. SEL 79-033, Stanford Univ.,Stanford Electronics Lab., Oct., 1979.

[22] J.L.Devore.
Probability&Statistics for Engineering and the Sciences.
Brooks/Cole Publishing Company, Monterey, California, 1982.

[23] S.W. Director and G.D. Hachtel.
The Simplicial Approximation Approach to Design Centering.
IEEE Trans. on Circuits and Systems CAS-24(7):363-372, July, 1977.

[24] S.W. Director, G.D. Hachtel, and L.M. Vidigal.
Computationally Efficient Yield Estimation Procedures Based on Simplicial Approximation.
IEEE Trans. on Circuits and Systems CAS-25(3):121130, March, 1978.

[25] D. A. Divekar.
DC statistical circuit analysis for bipolar IC's using parameter correlations - An experimental example.
IEEE Trans. CAD 3(1), Jan, 1984.

[26] R.W.Dutton, D.A.Divekar, A.G.Gonzalez, S.E.Hansen and D.A.Antoniadis.
Correlation of Fabrication Process and Electrical Device Parameter Variations.
IEEE Journal of Solid-State Circuits SC-12(5019-2):349-355, August, 1977.

[27] H. Eggleston.
: *Convexity.*
Cambridge University Press, London, England, 1958.

[28] N. J. Elias.
New statistical methods for assigning device tolerances.
In *Proc. of the 1975 ISCAS*, pages 329-332. IEEE, Apr., 1975.

[29] R.B.Fair and C.C.Tsai.
Profile Parameters of Implanted-Diffused Arsenic Layers in Silicon.
J. Electrochem. Soc. 123(4):583-585, April, 1976.

[30] W.Feller.
An Introduction to Probability Theory and Its Application.
John Wiley, 1968.

[31] A.V. Ferris-Prabhu.
Modeling of Critical Area in Yield Forecasts.
IEEE Journal of Solid-State Circuits SC-20(4):874-878, August, 1985.

[32] A.V. Ferris-Prabhu.
Defect Size Variations and Their Effect on the Critical Area of VLSI Devices.
IEEE Journal of Solid-State Circuits SC-20(4):878-880, August, 1985.

[33] R. Fletcher and M.J.D. Powell.
A Rapidly Convergent Descent Method for Minimization.
Computer Journal 6():163-168, 1963.

[34] S.K. Ghandhi.
VLSI Fabrication Principles.
John While&Sons, New York, 1983.

[35] D.J. Giannopoulos and S.W. Director.
IC Fabrication Process Optimization.
In *In Proc. of ICCAD - 84*, pages 164-166. Santa Clara, November, 1984.

[36] J. F. Gibbons, W. S. Johnson and S. W. Mylroie.
Projected Range Statistics.
Dowden, Hutchinson & Ross, Inc., Stroudsburg, Pennsylvania, 1975.

[37] G. H. Golub and C. F. VanLoan.
Johns Hopkins Series in the Mathematical Sciences: Matrix Computations.
Johns Hopkins University Press, Baltimore, 1983.

[38] A. Gupta, W.A. Porter, and J.W. Lathrop.
Defect Analysis and Yield Degradation of Integrated Circuits.
IEEE Journal of Solod-State Circuits SC-9(3):96-103, June, 1974.

[39] A. Gupta.
ACE: A Circuit Extractor.
In *In Proc. of the 20th Design automation Conference*, pages 721-725. June, 1983.

[40] G. D. Hachtel and S. W. Director.
A point basis for simplicial approximation.
In *Proc. of the Hull Conf. on Circuits and Systems.* Hull, England, 1977.

[41] W.E. Ham.
Yield-Area Analysis: Part I - A Diagnostic Tool for Fundamental Integrated-Circuit Process Problems.
RCA Review 39:230-249, June , 1978.

[42] H. H. Harman.
Modern Factor Analysis.
University of Chicago Press, Chicago, 1967.

[43] D. Heller.
Parallel Algorithms.
SIAM Review 20:740-773, 1978.

[44] R. B. Hitchcock Sr., G. L. Smith, D. D. Cheng.
Timing Analysis of Computer Hardware.
IBM Journal of Research and Development vol. 26(1), January, 1982.

[45] A.G.Ivakhnenko .
The Group Method of Data Handling, a Rival of the Method of Stochastic Approximation.
Soviet Automatic Control 13(3):43-55, 1968.

[46] R. A. Johnson and D. W. Wichern.
Applied Multivariate Statistical Analysis.
Prentice Hall, Inc., Englewood Cliffs, New Jersey 07632, 1982.

[47] P. Kager and A. J. Strojwas.
PI/C: Process Interpreter/Compiler.
In *IEEE International Conference on Computer-Aided Design (ICCAD) Digest of Technical Papers*, pages 321-323. IEEE, November, 1985.

[48] B.J. Karafin .
The General Component Tolerance Assignment Proglem in Electrical Networks.
PhD thesis, University of Pennsylvania, 1974.

[49] D.P. Kennedy, P.C. Murley, and R.R. O'Brien.
A Statistical Approach to the Design of Diffused Junction Transistors.
IBM Journal of Research and Development 8(5):482-495, November, 1964.

[50] C.S. Kim and W.E. Ham.
Yield-Area Analysis: Part II-Effects of Photomask Alignment Errors on Zero Yield Loci.
RCA Review 39:564-577, December, 1978.

[51] M. Kump and R.W. Dutton.
NATO ASI Series - No. 62: Two-Dimensional Process Simulation - SUPRA.
In " Process and Device Simulation for MOS-VLSI Circuits" edited by P. Antognetti et al., and published by Martinus Nijhoff Publishers, Boston, 1983.

[52] D. N. Lawley and A. E. Maxwell.
Factor Analysis as a Statistical Method.
American Elsevier Publ. Co., 1971.

[53] H.Lee, R.Dutton and D.Antoniadis.
On Redistribution of Boron During Thermal Oxidation of Silicon.
J. Electrochem. Soc. 126(11):2001-2007, November, 1979.

[54] E. L. Lehmann.
Testing Statistical Hypotheses.
Wiley, 1959.

[55] M.R. Lightner and S.W. Director.
Multiple Criterion Optimization with Yield Maximization.
IEEE Trans. on Circuits and Systems CAS-28(8):781-790, August, 1981.

[56] J. Linhard, M. Scharff, and H. Schiott.
Range Concepts and Heavy Ion Ranges.
Mat. Fys. Medd. Dan. Vid. Silsk. 33:1-39, 1963.

[57] K. K. Low and S. W. Director.
PED: A Graphical Process Editor.
In *Proceedings of the IEEE International Symposium on Circuits and Systems*, pages 560-566. IEEE, May, 1986.

[58] D. Luenberger.
Optimization by Vector Space Methods.
Wiley, New York, 1969.

[59] W.Maly.
A Method for Statistical Simulation of IC Manufacturing Process and Its Application in Process Control Problems, (in polish) .
In *Proc. of the Conference " Microelectronics76"*, pages 49-51. Torun, Poland, May, 1976.

[60] W.Maly and T.Gutt.
Base and Emitter Diffusion Simulation Model.
In *Proc. of International Conference on Computer Aided Design and Manufacturing of Electronics Components, Circuits and Systems*, pages 38-42. IEE, Sussex, August, 1979.

[61] W.Maly and A.J.Strojwas.
Simulation of Bipolar Elements for Statistical Circuit Design.
In *Proc. of ISCAS*, pages 788-791. Tokyo, July, 1979.

[62] W.Maly and S.W.Director.
Dimension Reduction Procedure for Simplicial Approximation Approach to Design Centering.
In *In Proc. of the 1980 European Conference on Circuit Theory and Design*, pages 115-120. Warsaw, 1980.

[63] W.Maly and S.W.Director.
Dimension Reduction Procedure for Simplicial Approximation Approach to Design Centering.
IEE Proc 127(6)(6):255-259, December, 1980.

[64] W.Maly, A.J.Strojwas and S.W.Director.
Fabrication Based Statistical Design of Monolithic IC's.
In *Proc. of ISCAS*, pages 135-138. Chicago, April, 1981.

[65] W. Maly and A.J. Strojwas.
Statistical Simulation of the IC Manufacturing Process.
IEEE Trans. on CAD of Integrated Circuits and Systems CAD-1(3):120-131, July, 1982.

[66] W.Maly and J.Deszczika.
Yield Estimation Model for VLSI Artwork Evaluation.
Electronics Letters 19(6):226-227, Mar., 1983.

[67] W.Maly.
Modeling of Point Defect Related Yield Losses for CAD of VLSI Circuits.
In *In Proc. of ICCAD - 84* , pages 161-163. Santa Clara, September, 1984.

[68] W.Maly.
Modeling of Random Phenomena in CAD of IC - A Basic Theoretical Consideration.
In *Proc. of ISCAS 85*, pages 427-430. Kyoto, 1985.

[69] W.Maly.
Modeling of Lithography Related Yield Losses for CAD of VLSI Circuits.
IEEE Trans. on Computer-Aided Design of Integrated Circuits and Systems CAD-4(4):166-177, July, 1985.

[70] W.Maly, A.J. Strojwas, and S.W. Director.
VLSI Yield Prediction and Estimation: A Unified Framework.
Research Report CMUCAD-85-67, SRC-CMU Center for Computer-Aided Design, Department of Electrical and Computer Engineering, Carnegie-Mellon University, October, 1985.

[71] W. Maly, B. Trifilo, R. Hughes, and A. Miler .
Yield Diagnosis Through Interpretation of Tester Data .
In *In Proc. International Test Conference*, pages 10-20. September, 1987 .

[72] W. Maly, M.E Thomas, J.Chinn and D. Campbell.
Characterization of Type, Size and Density of Spot Defects in the Metalization Layer (edited by W.Moore, W. Maly and A.J. Strojwas).
Yield Modeling and Fault Tolerance in VLSI :72-90, 1988.

[73] T.E.Mangir.
Sources of Failures and Yield Improvement for VLSI and Restructurable Interconnects for RVLSI and WSI: Part I - Source of Failures and Yield Improvements for VLSI.
Proc. of IEEE 72(6):690-708, June, 1984.

[74] R. S. Martin and J. H. Wilkinson.
Similarity Reduction of a General Matrix to Hessenberg Form.
Numerische Mathematik 12:349-368, 1968.

[75] M. D. Matson.
Macromodeling of Digital MOS VLSI circuits.
In *Proceedings of the 22nd Design Automation Conference.* 1985.

[76] G.E. Moore.
A Simple Method for Modeling VLSI Yields.
Electronics 43:126-130, 1970.

[77] P. K. Mozumder.
Statistical Quality Control of VLSI Fabrication Processes.
Master's thesis, Carnegie-Mellon University, March, 1986.

[78] P.K. Mozumder, A.J. Strojwas and D. Bell.
Statistical Process Simulation for CAD/CAM.
In *Proceedings of 1988 CICC*, pages 13.5.1-13.5.4. 1988.

[79] B.T. Murphy.
Cost-Size optima of Monolithic Integrated Circuits.
Proc. IEEE 52:1537-1545, December, 1964.

[80] L.W. Nagel.
SPICE2: A Computer Program to Simulate Semiconductor Circuits.
Memorandum ERL-M520, Electronics Research Laboratory, College of Engineering, University of California, Berkeley, May, 1975.

[81] S. R. Nassif, A. J. Strojwas, and S. W. Director.
FABRICS II - A Statistical Simulator of the IC Fabrication Process.
Proceedings of the IEEE Intl. Conference on Circuits and Systems :298-301, September-October, 1982.

[82] S. R. Nassif, A. J. Strojwas and S. W. Director.
FABRICS II, A Statistical Simulator of the IC Fabrication Process: User's Manual
SRC-CMU Center for Computer Aided Design, 1983.

[83] S.R.Nassif,A.J.Strojwas and S.W.Director.
FABRICS II: A Statistically Based IC Fabrication Process Simulator.
IEEE Trans. on Computer-Aided Design of Integrated Circuits and Systems CAD-3(1):40-46, January, 1984.

[84] S. R. Nassif, A. J. Strojwas and S. W. Director.
A Methodology for Worst-Case Analysis of Integrated Circuits.
IEEE Transactions on Computer-Aided Design of Integrated Circuits and Systems CAD-5(1):104-113, January, 1986.

[85] T. Okabe et al.
A Simple Method for Modeling VLSI Yields.
Elec. Eng. in Japan 92:135-141, 1972.

[86] A. Papoulis.
Probability, Random Variables, and Stochastic Processes.
McGraw-Hill Book Company, 1965.

[87] O. Paz and T. Lawson,Jr.
Modification of Poisson Statistics: Modeling Defects Induced by Diffusion.
IEEE Journal of Solod-State Circuits SC-12(5):540-546, October, 1977.

[88] J. R. Pfiester, .
PRIDE - Portable Process and Device Design
Integrated Circuits Laboratory,Stanford University, 1981.

[89] J.F. Pinel and K.A. Roberts.
Tolerance Assignment in Linear Networks using Nonlinear Programming.
IEEE Transactions on Circuit Theory CT-19():475-479, 1972.

[90] J.F. Pinel and K. Singhal.
Efficient Monte Carlo Computation of Circuit Yield Using Importance Sampling.
In *Proc. of ISCAS 77*, pages 575-578. 1977.

[91] E. Polak and A. Sangiovanni-Vincentelli.
Theoretical and Computational Aspects of the Optimal Design Centering, Tolerancing, and Tuning Problem.
IEEE Trans. on Circuits and Systems CAS-26(9):795-813, September, 1979.

[92] M.J.D. Powell.
A Fast Algorithm For Nonlinearly Constrained Optimization Calculations.
Unpublished , June, 1977.
Presented at the 1977 Dundee Conference on Numerical Analysis.

[93] N. B. Rabbat, W. D. Ryan, and Q. A. M. A. Hossain.
A Computer Modeling Approach for LSI Digital Structures.
IEEE Transaction on Electron Devices vol. ED-22(8), August, 1975.

[94] P.J.Rankin.
Statistical Modeling for Integrated Circuits.
IEE Proc. - G 129(4):186-191, August, 1982.

[95] C. R. Rao.
Linear Statistical Inference and its Applications.
Wiley, New York, 1973.
2nd. Edition.

[96] K.V. Ravi.
Imperfections and Impurities in Semiconductor Silicon.
John Wiley&Sons, New York, 1981.

[97] R. Razdan and A.J. Strojwas.
Statistical Design Rule Developer.
In *Proc. of ICCAD 85*. Santa Clara, 1985.

[98] D. Riley and A. Sangiovanni-Vincentelli.
Models for a New Profit-Based Methodology for Statistical Design of Integrated Circuits.
IEEE Transactions on Computer-Aided Design CAD-5(1), January, 1986.

[99] M. J. Saccamango.
Efficient Modeling of Small-Geometry MOSFET's.
PhD thesis, Carnegie Mellon University, Department of Electrical and Computer Engineering, April, 1987.

[100] S. Selberherr, W. Fichtner, and H.W. Potzl.
MINIMOS - A Program Package to Facilitate MOS Design and Analysis.
In Numerical Analysis of Semiconductor Devices, by B.T. Browne and J.J.H. Miller, Eds., Dublin, Ireland: Bool Press , 1979.

[101] J. H. Shelly and D. R. Tryon.
Statistical Techniques of Timing Verification.
In *Proceedings of the 20th Desing Automation Conference.* 1983.

[102] C.R. Shyamsundar.
Adaptive Control of VLSI Fabrication Processes.
Master's thesis, Carnegie-Mellon University, March, 1986.

[103] K. Singhal and J. F. Pinel.
Statistical Design Centering and Tolerancing Using Parametric Sampling.
In *Proceedings of ISCAS 80.* 1980.

[104] C.J.Spanos and S.W.Director.
PROMETHEUS: A Program for VLSI Process Parameters Extraction.
In *In Proc. of ICCAD - 83*, pages 176-177. Santa Clara, September, 1983.

[105] C.J. Spanos and S.W. Director.
Parameter Extraction for Statistical Process Characterizatio.
IEEE Trans. on Computer Aided Design (1):66-78, January, 1986

[106] C.H. Stapper.
Defect Density Distribution for LSI Yield Calculations.
IEEE Trans. on Electron Devices ED-20:655-657, July, 1973.

[107] C.H. Stapper.
On a Composite Model to the IC Yield Problem.
IEEE Journal on Solid-State Circuits SC-10:537-539, December, 1975.

[108] C.H. Stapper.
LSI Yield Modeling and Process Monitoring.
IBM Journal of Research and Development 20(3)():228-234, 1976.

[109] C.H. Stapper et al.
Yield Modeling for Productivity Optimization of VLSI Memory Chips with Redundancy and Partially Good Product.
IBM J. Res. Develop. 24(3):398-409, May, 1980.

[110] C. H. Stapper.
Yield Model for 256K RAMs and Beyond.
In *Proc. of 1982 International Solid-State Circuits Conference*, pages 12-13. 1982.

[111] C. H. Stapper, R. M. Amstrong and K. Saji.
Integrated Circuit Yield Statistics.
Proc. of the IEEE 71(4):453-470, April, 1983.

[112] C.H.Stapper .
Modeling of Integrated Circuit Defect Sensitivities .
IBM J. Res. Develop. 27 (6):549-557, November, 1983.

[113] M. L. Stein.
An Efficient Method of sampling for statistical Circuit Design.
IEEE Transactions on Computer-Aided Design of Integrated Circuit and Systems CAD-5(1):23-29, January, 1986.

[114] J. Stoer and C. Witzgall.
Convexity and Optimization in Finite Dimensions.
Springer, New York, 1970.

[115] A.J.Strojwas, S.R.Nassif and S.W.Director.
A Methodology for Worst Case Design of Integrated Circuits.
In *In Proc. of ICCAD - 83*, pages 152-153. Santa Clara, September, 1983.

[116] M.A. Styblinski and L.J. Opalski.
Random Perturbation Method for IC Yield Optimization with Deterministic Process Parameters.
In *Proc. of ICCAD-84*, pages 977-980. Santa Clara, November, 1984.

[117] M.A. Styblinski and L.J. Opalski.
Algorithms and Software Tools of IC Yield Optimization Based on Fundamental Fabrication Parameters.
IEEE Transaction on Computer-aided Design CAD-5(1):79-89, January, 1986.

[118] S.M. Sze.
VLSI Technology.
McGraw-Hill Book Company, New York, 1983.

[119] G. Taguchi and M. S. Phadke.
Quality Engineering Through Design Optimization.
IEEE Communication Society (Global Telecommunications Conference) , November, 1984.

[120] K.S. Tahim and R. Spence.
A Radial Exploration Approach to Manufacturing Yield Estimation and Design Centering.
IEEE Trans. on Circuits and Systems CAS-26(9):768-774, September, 1979.

[121] A.R. Thorbjornsen and S.W. Director.
Computer-aided Tolerance Assignment for Linear Circuits with Correlated Elements.
IEEE Transactions on Computer-Aided Design CT-20():518-524, 1973.

[122] M. Trick, A.J. Strojwas, and S.W.Director.
Fast RC Simulation for VLSI Interconnect.
In *Proc. of the ICCAD83*, pages 178-179. 1983.

[123] L.M. Vidigal and S.R. Nassif and S.W. Director.
Cinnamon: Coupled Integration and Nodal Analysis of MOS Networks.
In *Design Automation Conference Proceedings*. IEEE, July, 1986.

[124] S.Wagner.
Diffusion of Boron from Shallow Ion Implants in Silicon.
J. Electrochem. Soc. 119(11):1570-1576, November, 1972.

[125] H.Walker and S.W.Director.
Yield Simulation for Integrated Circuits.
In *Proc. of ICCAD-83*, pages 256-257. IEEE, Santa Clara, Sept., 1983.

[126] H. Walker and S. W. Director.
VLASIC: A Catastrophic Fault Yield Simulator for Integrated Circuits.
IEEE Trans. on CAD of Integrated Circuits and Systems CAD-5(4):541-556, October, 1986.

[127] H. Walker.
Yield Simulation for Integrated Circuits.
Kluwer Academic Publishers, Boston, 1987.

[128] D. E. Wallace and C. H. Sequin.
Plug-In Timing Models for An Abstract Timing Verifier.
In *Proceedings of the 23rd Design Automation Conference.* 1986.

[129] R.M. Warner,Jr.
Applying a Composite Model to the IC Yield Problem.
IEEE Journal on Solid-State Circuits SC-9:86-95, June, 1974.

[130] C.P.Wu, E.C.Douglas and C.W.Mueller.
Redistribution of Ion-Implanted Impurities in Silicon During Diffusion in Oxidizing Ambients.
IEEE Trans. Elect. Dev. :1095-1097, September, 1976.

[131] T.Yanagawa.
Yield Degradation of Integrated Circuits Due To Spot Defects.
IEEE Transactions on Electron Devices ED-19(2)(2):190-197, 1972.

[132] J.F.Ziegler, B.L.Crowder, G.W.Cole, J.E.E. Baglin, and B.J.Masters.
Boron Atom Distributions in Ion-Implanted Silicon by the $(n,{}^4\mathrm{He})$ Nuclear Reaction.
Appl. Phys. Lett. 21(1):16-17, July, 1972.

Index

2-norm 63

Acceptability region 6, 26, 27, 28, 242
Accuracy 96

Bird's beak effect 214
Box constraints 45
Break 186, 189, 193, 215

Caltech Intermediate Form 111
Catastrophic yield estimation 2
Clustering 178, 180, 184, 225
Clustering coefficient 178
Combined deformations 219
Computer aided manufacturing 229
Contamination 176
Control 3
Cost of manufacturing 37, 38
Critical area 30, 190, 198, 222
Cross talk 214
Crystal dislocation 176

Defect densities 184
Defect radii 184, 186
Defect size 177
Defect spatial distribution 178
Dependency trees 247
Depletion layer expansion 214
Deposition 117
Design centering 1, 43, 46
Design centering procedure 51
Design for manufacturability 231
Design rules 200
Design yield 27, 28
Designable parameter 168, 166, 167, 168, 171, 172
Designable random variable 88, 88
Designable variable 166, 167
Device model library 119
Device parameter 152, 159, 160, 161, 162, 164, 165
Differential equations 96
Diffusion 109
Dimension reduction 84
Dimension reduction problem 85
Distance from a point to a hyperplane 65

Disturbances 3, 95
Disturbed timing 136
Double-Bridge test structure 186
Drive-in 109, 117
Dry etching 117
Dual norm 64

Electrical deformations 18
Epitaxy 117
Equivalent layout 224
Extraction stage 116

Fabrication cost minimization 37
FABRICS 41, 93, 100, 105, 106, 108, 109, 112, 116, 117, 124, 126, 129, 152, 155, 156, 160, 161, 166, 167, 168, 169, 171, 173, 238, 240, 244, 261, 264
Fault analysis 205
Faults 21, 176
Feed forward control 235
Final characterization phase 115
Final testing yield 25, 31
Final tests 6
Fisher's linear discriminant 255
Front end costs 230
Functional acceptability region 29
Functional performance 26
Functional test and measurements 10
Functional test thresholds 10
Functional tests 4
Functional yield 29
Functional yield losses 175
Functionally acceptable performances 29

Gaussian distribution 195
Geometry deformations 13
Global deformations 20, 219
Global disturbances 20
Global electrical deformations 18

Hard-performance faults 21

IC probe measurement 6, 9
IC probe selection 6
IC selection thresholds 9
Ideal IC 13
Importance sampling 151, 130, 151
In-line measurements 3, 8
Independently designable random variable 88, 89
Information flow 3, 4

Inspection step 3
Instabilities of process conditions 11
Inter-die 98
Inter-die variation 131, 152, 167
Interlayer failures 223
Intra-die 98
Intra-die variation 131, 152, 153, 167
Ion implantation 106, 110

Joint failures 213, 219
Jpdf-norm 64
Junction leakage 206, 219

Lateral diffusion 214
Lateral edge displacement 13, 15
Lateral interlayer failures 216
Layout 3, 94
Leakage 176
Line registration 195
Line registration error 195, 197
Line registration errors 188
Linewidth variation 214
Lithography defects 177
Lithography spots 12
Local deformations 20
Local disturbances 20
Local electrical deformations 19
Local process disturbances 175
Logic model 136, 143
Logic simulator 129

Macrocell 218, 220, 226
Mahalanobis distance 255
Manufacturing operation 3
Manufacturing process 3
Manufacturing yield 22, 23, 31
Mask 3
Mask editor 119
Mask misalignment 15, 197
MINIMOS 105
Misalignment 16, 196, 216, 218
Misalignment vectors 196
Missing gate 207
Mixed worst-case 62
Mixture 100
Monte Carlo approach 203
Monte Carlo technique 176
Monte Carlo yield simulation 205

Multi-terminal transistor 211, 210
Multilevel structure 99
Multiple transistor defect 207

Negative binomial 177
New device 209, 210
Nominal IC layout 199
Nominal simulation mode 124
Nominal timing 136, 137, 147
Nominal values 114
Norm 62
Norm body 64

Off-line control 231
Open 206
Open device 207
Open via 206
Open wire 206
Oxidation 117
Oxide pinhole 206
Oxide pinholes 176

Parameterized cell 166
Parametric acceptability region 29
Parametric fault 137, 150
Parametric performance 26
Parametric tests 4
Parametric yield 29
Parametric yield optimization 1
Parametrically acceptable performances 29
Performance faults 21
Performance space 27
PEW system 117
Photodefects 177
Photolithography 110
Pinhole 176
Pinholes 218
PISCES 117
Point defects 12, 175
Poisson 191
Poisson distribution 178
Predeposition 109, 117
Pretuning phase 115
Primary design variables 87
Primitive failures 213, 214
Principal component analysis 82
Probabilities of failure (POF) 213
Probe testing yield 25, 31

Probe tests 4
Probe yield 24
Process characterization problem 112
Process control 94
Process controlling parameters 3
Process disturbance 11, 94, 97, 98, 135, 155, 156, 158, 160, 161, 162, 164, 165, 167
Process efficiency 22
Process Engineer's Workbench 116, 117
Process Interpreter 121
Process model library 117
Process monitoring 231
Process state after assembly and packaging 10
Process state after final testing 10
Process state after IC selection 10
Process state variables 6
Process state vector 9
Process supervisor 117
Process trajectory 233
Processing yield 25
Processor interpreter/compiler 117
PROMETHEUS 93, 100
Punch through 214

Quality control 231
Quality improvement 232
Quantity control 231

Radial distribution 178
Random defects 175
Random number generators 204, 99
Random phenomena 11
Rayleigh distribution 185
Regression model 167
Revenue 237
Rework 235

Scaling factor 202
SEDAN 105
Selection region 6
Selection step 3
Selection thresholds 3
Sensitive area 222
Short 186, 189, 193, 205
Shorted device 207
Shorts 214
Silicon surface defects 19
Simplicial approximation 43, 46, 248
Simplicial approximation method 83, 86

Soft-performance faults 21
Spatial distribution 177
Spikes 19
Spot defects 17, 175
Stacking of layers 121
State of an IC chip 7
Statistical model 130, 135, 136, 138, 147
Statistical process simulation 2, 93
Statistical process simulator 167
Structural faults 21
Structure of FABRICS 93
Structure of the process 4
SUPREM 105, 117
Surface treatment processes 110

Test chips 3
Test conditions 3, 9
Test structure 3, 4
Test structure measurements 9
Test structure test conditions 9
Test structure thresholds 9
Thermal oxidation 110
Three-layer failure 216
Threshold of acceptability 234
Threshold of acceptance 256
Throughput 229
Timing model 147, 131, 134, 135, 136, 137, 139, 143, 147
Timing verifier 130, 131, 134, 152
Tolerance assignment 1
Tolerance assignment problem 76
Tuning 93, 112
Two-level Gaussian random vector 103
Two-level random variable 100

Verification stage 116
Vertical effects 16
Vertical interlayer failures 218
Virtual layout 176, 198, 199, 224
VLASIC 203, 204, 212

Wafer map 204
Wafer yield 24, 31
Worst-case 129, 152, 153, 154, 155, 156, 158, 160, 161, 164, 165
Worst-case parameters 70
Worst-case tolerance assignment 62

Yield 1, 46
Yield enhancement 265

Yield estimation 23
Yield loss 198
Yield losses 191
Yield maximization 62, 93
Yield maximization problem 83
Yield modeling 175
Yield prediction 26